Balloonists, Alchemists, and Astrologers of the Nineteenth Century: The Tale of George and Margaret Graham

Daniel Harms

Cover image, "New Hungerford Market, London, on the Day of Opening, July 2, 1833—with Ascent of Mr. Graham in his Balloon," courtesy Yale Center of British Art, Paul Mellon Collection.

First Edition
27 26 25 24 23 22 21 20 19   2 3 4 5

Contents

Introduction

Graham first caught my attention when I found a mention of a "Key of Solomon" manuscript at the Cleveland Public Library. I knew that any manuscript with that title could be any sort of work, so I stopped there on a day-long drive from Upstate New York to Michigan to view it. The manuscript intrigued me, as did the George Graham it was intended for.

No one has treated George and Margaret Graham at any length. Scholars of esotericism make an occasional brief note, and references to the pair in several archives in the US and UK remain untouched. Historians of ballooning are certainly aware of the Grahams, but aside from their reputation for crashes, they form little more than a footnote in the narrative of the triumph of the skies. Although they were celebrities of their time, George and Margaret are practically forgotten.

Taking my inspiration from Ellic Howe's pamphlet on the Grahams' contemporary Raphael, I will be presenting what I have learned about these two individuals. I hope this will be not only a fascinating account of their lives, but a spur to future scholarship into both aerostation and occultism in the nineteenth century.

Acknowledgements

Thanks to Robert Derie, Joscelyn Godwin, and Mary Ruwell and the staff at the United States Air Force Academy Library, Crestina Signorelli at the Wellcome Institute, Brian Riddle at the National Aeronautical Library, and the staff at the British Library, the National Archives at Kew, the University College of London, the Folger Shakespeare Library, the Senate House Library, and the City of Westminster Archives.

Special thanks to Pauline Alston and Ben Fernee for all their assistance. Thanks to Casey Hickey for his layout skills and invaluable advice.

'Very Aspiring Ideas': Birth and the Early Geography of George Graham

George William Graham was born on November 13, 1784, 10:10 PM, in St. James Parish, London, to William and Rebecca Graham. He was baptized on December 26, 1784 at St. Margaret Westminster. His friend and astrological colleague Raphael would later provide the following judgment on the timing on George's birth:

> The powerful and regal sign Leo is in the ascendant, and the qualities of his mind, his firmness, and strength of nerve, and his singular courage, are clearly shown, by his being born under the *solar* influence; by the conjunction of **Mars**, **Sol**, and **Mercury**, in a *fixed* sign, and the whole of these planets applying to **Jupiter**; also in a sign of *fixed* nature, by a square (and determined) aspect, from powerful angles and constellations of peculiar influence over the fate of mortals.
>
> The conjunction of **Mars** with **Mercury**, and the **moon** with **Venus**, denote a surprising degree of mechanical genius and inventive ideas, which the native is well known to possess; while the *trine* of **Herschell** to each of these significators sways the mental faculties to pursuits of no common or ordinary kind, but mostly out of the reach of custom, and those remarkable for strangeness or eccentricity. It also gives very aspiring ideas.[1]

We know very little of George's family, or his early life. His common family name and the fragmentary nature of the records from the period make finding information difficult. I cannot confirm, for instance, the statement that his father might have been born in 1760. It might be that those with greater genealogical acumen will be able to fill in important aspects of his childhood. What we do know, however, is the place in which he lived.

At an age of approximately nine or ten, George (and presumably his family) moved into the house at 41 Poland Street, Soho, where

---

1. Raphael and Anglicus, *The Astrologer of the Nineteenth Century*, 440.

he would dwell for much of his life. Soho had been built by nobles and rich merchants attempting to escape the city for more pleasant surroundings, but by the 1790s they had fled once again. Many of the houses were subdivided into cheap tenements, and all manner of vices could be encountered in the streets. Nonetheless, immigrants coming to London often found a new life and cheap lodgings here, and Soho was a popular place with artists, musicians, and craftsmen.[2]

Poland Street, on which the Grahams lived, was formed out of land occupied by a horse pasture called Little Gelding's Close. When the land was designated as a street in 1689, it took its name from the King of Poland Tavern across the street. The first houses were built in 1705, with the street filling up in two years' time. Stepping out on Poland Street, George could have seen buildings that were, for short periods, home of such luminaries as William Blake and Percy Bysshe Shelley, as well as the King's Arms Tavern, where the Ancient Order of Druids held its first meeting.[3]

The most notable structure in the neighbourhood, with which the Graham residence shared a garden wall, was the Pantheon. Originally opened in 1772, the first building of that title was an entertainment palace intended to be so prestigious that no one was admitted without a recommendation from a peeress of the realm. As popular fashions changed, the place was remade into an opera house, which burned down on January 14, 1792. By the time the Grahams would have arrived, the grounds would have been occupied by a Pantheon rebuilt by one Crispus Claggett, who held concerts and masquerades on that site until he vanished some years thereafter. (According to author James Winston, George later claimed that a dream revealed to him a set of documents, presumably having to do with Claggett's disappearance, beneath the stage.) The venue languished until 1814, when a dispute with the Lord Chamberlain caused its doors to be closed for the following twenty years. It is best known to occultists due to the summoning of elemental forces that Éliphas Lévi and Edward Bulwer-Lytton supposedly performed upon its roof, according to Arthur Edward Waite. The

2. Great Britain Central Criminal Court and Buckler, *Central Criminal Court. Minutes of Evidence*; Summers, *Soho: A History of London's Most Colourful Neighbourhood*.

3. London City Council, *Survey of London*, 31–32:243–49.

Pantheon would be an important site in George's life, and the mix of mysticism, theatricality, and mystery that surrounds the building is emblematic of that life.[4]

'Witness to Some Very Curious Experiments': George as Occultist, The Early Days

For considerable years before he became a celebrity, George is reported to have pursued mystical research. In addition to the above statements from George's nativity, Raphael notes that the man's chart shows a propensity to *"abstruse* studies," borne out in how he had "gone very considerable lengths in occult philosophy, particularly the *alchemical art and transmutation of metals*, in which science I have been witness to some very curious experiments of his performing, and for the effects of which a modern chemist would find it difficult to assign the cause." According to Schuchard, George became the central figure of his own occult circle in 1818. Not only does she not give the source for this information, however, she makes the erroneous statement that his "death" in an 1824 balloon accident brought an end to that group, raising questions as to how reliable the evidence of his magical group might be.[5]

It was about the same time that Raphael, or Robert Cross Smith, made his way into George's life. According to Ellic Howe, Smith had recently arrived from Bristol and was likely working as a clerk on Upper Thames Street. After being introduced to George's associates, however, Smith decided to turn astrology into a professional career. He moved to No. 5, Castle-street East (modern Eastcastle Street), and lived for some time with George's financial assistance, according to the astrologer 'Dixon.' Raphael and Mr. Graham shared other interests, including alchemy and ritual magic.[6]

---

4. London City Council, 31–32:268–83; Lévi and Waite, *The Mysteries of Magic: A Digest of the Writings by Éliphas Lévi*, 8.

5. Raphael and Anglicus, *The Astrologer of the Nineteenth Century*, 441; Schuchard, *Freemasonry, Secret Societies, and the Continuity of the Occult Traditions in English Language*, 486.

6. Howe, *Raphael; or, The Royal Merlin*, 12–13.

As the dream revelation regarding Claggett shows, George was no stranger to uncanny experiences. He told a 'Friend' (most likely Raphael) another tale regarding the Pantheon:

"On the 22d of June, 1821, in the evening, Mr. G. called upon me in great agitation, and gave me an account of a most extraordinary supernatural occurrence respecting the appearance of two spirits, which he had just seen, as follows:—Late in the evening of this day, himself and three friends were going into the P_____n, Oxford Street, to have a view of some philosophical process relative to aerostation, which was there being pursued, when, as he advanced to the edge of the pit, he was surprised to see the appearance of two boys, who were perfectly black in visage and clothing; they stood at the end of the entrance hall, where was a chasm of nearly nine feet in depth from the dilapidated state of the building, and into which, most probably, the whole company would have fallen, had not these visionary beings appeared in that place, as if blocking up the road thereto. When the company came within seven or eight yards distance of them, they suddenly turned round, and instantaneously disappeared. Four distinct sounds were heard, and no more. Now it was utterly impossible that any *human* beings could either secrete themselves or run away, for the staircase was nearly thirty yards distant, and Mr. G. sent his companions immediately in every direction by which they could escape, but no one was to be found, added to which there was no possible outlet, the doors were all locked, and the windows closed, so that no human being could have escaped, unless they either sunk into the earth, or took their flight through the roof of the building." This is verbatim as the aeronaut related it.[7]

The dilapidated state of the building might be a result of the auction when the Pantheon was closed in 1814, in which most of the interior trappings, including the floor of the orchestra pit, were sold. We cannot be sure of what sort of "philosophical process" George might have been on his way to witness. The aeronaut Charles Green made his first

7. Raphael and Anglicus, *The Astrologer of the Nineteenth Century*, 512.

flight, also the first to use coal gas, on July 19, so he could have stored his balloon in the building.[8]

George's mystical interests and friendship with Raphael brought about a collaborative publication—and his only book—in 1822. It bore the weighty title of *The Philosophical Merlin, being the Translation of a Valuable Manuscript, Formerly in the Possession of Napoleon Bonaparte: Found amongst Other Valuable Papers, in his Cabinet, at the Battle of Leipsic: Known to have been Highly Prized by Him, and Consulted for the Choice of His Generals; Exhibiting a Curious Specimen of the Ancient Mystical Learning, Practised by the Orientals; and Affording the Most Scientific & Rational Amusement; by Enabling the Reader to Cast the Nativity of Himself, or Any Other Person, without the Aid of Tables, Instruments, or Arithmetical Calculation; and in a Few Minutes, to Ascertain Every Particular relaling [sic] to Their Future Fate; by the Rules of the Ancient Geomancy.* The book was attributed to "R. C. S., Philo. Astr." and "G. W. G., F. R. C." and published by John Denley on Catherine Street.

The English regarded Napoleon with fear and dread, so it is unsurprising that Graham and Raphael summoned up the memory of the recently-deceased general to lend their project the appropriate air. Prussian soldiers supposedly obtained the book at the Battle of Leipsic (October 16–19, 1813), a decisive defeat of the Emperor by combined Prussian and Russian forces. The emperor's forces left behind 800 wagons taken by the enemy, so this might have been seen as a plausible story, at least for a contemporary audience. This scenario is possible, but given the general history of pseudonymous occult manuscripts with fabricated histories, it is best to regard it with scepticism.[9]

The Philosophical Merlin was dedicated on August, 1822 on the day of Mercury (Wednesday) to Marie Anne Lenormand (1772–1843). Mlle. Lenormand was a famous French fortune-teller who spent much of her life in Paris. Much of what is known about her is blatantly inaccurate, but she most likely had clients among the upper classes of France, although her ties to the Empress Josephine might be exaggerated. As we shall see, rumour has it that she made an extensive trip to London and contacted Raphael.

8. London City Council, *Survey of London*, 31–32:268–93.

9. Smith, *1813, Leipzig: Napoleon and the Battle of the Nations*, 297.

As for the book's title, Merlin was a popular figure in British discourse, in a wide variety of different roles, including that of prophet. In 1644, William Lilly published a collection of prophecies as *Merlinus Anglicus Junior*. When this proved lucrative, he followed it three years later with his almanac *Merlini Anglici Ephemeris*. 1656 saw the release of the *British Merlin*, another series of almanacs, from Schardanus (or Cardanus) Rider. Both almanacs were still published and popular in 1813, so it is not surprising that Graham and Smith sought to bring themselves success through association with these works.[10]

If we compare the system in *The Philosophical Merlin* to the most popular work on the topic—the treatise "Of Geomancy," attributed to Heinrich Cornelius Agrippa and included in the spurious *Fourth Book* attributed to him—we find that the book's system is vastly simplified. Both systems require the creation of a series of lines of marks which are then counted to create one of sixteen figures. In pseudo-Agrippa's system, sixteen lines resolve into four figures, which are then used to create a total of fifteen figures to be interpreted in conjunction with each other. *The Philosophical Merlin* simply adds together two figures to create one, which is then taken as absolutely determinative of an individual's personality, down to occupation, marriage, and the position of moles. Raphael and Graham assure readers that the system works if the person makes no mistakes and is serious. Those who do not find the results to be accurate, however, should try again in an hour or so. In his work on the history of geomancy, Stephen Skinner observes that *The Philosophical Merlin* must be derived from some unknown work on astrogeomancy.[11]

At the end of the work, the authors promised that a second part, including "a number of curious and valuable Secrets in ASTROLOGY AND GEOMANCY," would appear. It was never published. In the *True Prophetic Messenger*, published in 1833, the year following Raphael's death, "Dixon" mentioned the *Merlin*, stating that "whatever may be

10. Knight, *Merlin: Knowledge and Power through the Ages*, 117.

11. Agrippa von Nettesheim and Petrus, *Fourth Book*, 1–29; G. and S., *The Philosophical Merlin*; Skinner, *Geomancy in Theory and Practice: The Most Complete History of Western Divinatory Geomancy in English*, 161.

its merit, it did not succeed, and had but a very limited sale, and was eventually sold off to Mr. Denley."[12]

George Graham, the Rosicrucians, and the Mercurii

Did George's mystical friendships turn into an esoteric group? George declared himself a member of one such organization, the Rosicrucians, and was closely associated with the members of another, the Mercurii.

By including the initials "F. R. C." in *The Philosophical Merlin*, Graham declared himself as a member of the "Fraternitas Rosae Crucis," the Brotherhood of the Rosy Cross, better known as the Rosicrucians. The Rosicrucian movement's story begins in the early seventeenth century, when three manifestos were published detailing its supposed history and beliefs. Such beliefs had become fashionable in esoteric circles by the late eighteenth century, with many individuals declaring themselves to be Rosicrucians or seeking membership in the order. By the early nineteenth century, high-degree Masonic degrees based upon the symbolism of Rose Croix had filtered into English lodges, possibly via Mason involved with Templar rites, or through interactions with French and Irish lodges and Masons. Still, we know of no Masonic groups in which George was involved, and he makes no statements as to how he interprets the meaning of Rosicrucianism which would allow him to be placed with one group or another.[13]

What "Rosicrucians" would have been active in George's time and place? Making these determinations should be pursued with caution. For example, some claimed that the celebrated Jewish mystic and Baal Shem Doctor Hayyim Falk had been involved in Rosicrucianism, although he passed away two years before George's birth. Nonetheless, we have three figures active at the time with similar interests to his,

---

12. Dixon, *The True Prophetic Messenger for 1833: Containing Remarkable Events, Predictions, the Weather, &c. ...*, 86.

13. McIntosh, *The Rosicrucians*; Jackson, *Rose Croix: The History of the Ancient and Accepted Rite for England and Wales*, 88–89, 96–109.

who could be potential Rosicrucian initiators or colleagues.[14]

The first individual with whom George might have had Rosicrucian ties was Sigismund Bacstrom (c. 1740–c. 1808). Bacstrom claimed to be born in Germany and educated at the University of Strasburg, although no record of this exists. He later made London his home but served as a secretary and ship's doctor on many scientific and trading expeditions across the world. He often found the conditions insufferable and moved from one ship to another. Bacstrom was often in financial difficulty, and his fate remains a mystery.

Bacstrom is known today for two chief accomplishments. First, he made a number of technically competent illustrations of Native Americans of the northwest coast of North America of use to anthropologists and historians. Second, he was an advocate of mystical practices, including alchemy and Rosicrucianism.

During one of Bacstrom's trips, war broke out between England and France, endangering the shipping lanes. Bacstrom was engaged to the ship *Warren Hastings*, the captain of which made the poor decision to hire a French first mate and crew loyal to him. The mate led a mutiny and sailed the ship to the French colony of Mauritius. It was there that Bacstrom supposedly met the mysterious Comte de Chazal, who initiated him into the brotherhood of the Rosicrucians. As Bacstrom tells it, the Comte was not only incredibly wealthy, but could also make gold in his private laboratory via alchemy. The association was brief, for after only a few months Bacstrom was ordered onto a ship to New York City.[15]

We lack information as to how much Bacstrom promoted Rosicrucianism upon his return to London. Based upon Bacstrom's charter and the one he gave to Alexander Tilloch on April 5, 1797, the brotherhood had at least two ranks, the Member Apprentice and the Practical Member and Brother. As each Practical Member might only initiate

---

14. Westcott's claim that William Henry White, Grand Secretary of English Freemasonry, was initiating individuals into Rosicrucianism has been debunked, as the document referring thereto actually refers to the Order of the Red Cross of Constantine. Wilson, "The Origin of Our Rosicrucian Society," 10.

15. Cole, "Sigismund Bacstrom's Northwest Coast Drawings and an Account of His Curious Career"; McIntosh, *The Rosicrucians*, 92; Waite, *The Brotherhood of the Rosy Cross, Being Records of the House of the Holy Spirit in Its Inward and Outward History*, 558.

up to two Member Apprentices, the growth of any such order must have been limited.[16]

The other aspect of Bacstrom's esotericism, that of alchemy, brought him in touch with many more people. As early as 1786, Bacstrom managed to acquire a position as an assistant in chemical experimentation, based upon his purported ability to employ magnets to aid plant growth. When his patron passed away in 1789, the alchemist Peter Woulfe offered two pounds toward an abortive subscription for his welfare. After his return to London, Bacstrom assembled his own corresponding group of alchemists, of whom we know only a few—a Mr. Ford, a Mr. Hand, and a Captain Alefson, the latter of whom nearly poisoned himself, Bacstrom, and his wife through poisonous gas through his intemperate action during an experiment. Perhaps more names lie in the thousands of pages of Bacstrom's alchemical documentation recently uploaded by the Getty Research Library, and these might help to illuminate both his experimental colleagues and those connected with his other interests.[17]

What makes this particularly intriguing is that one of George's surviving treatises on alchemy is bound with three holographs by Bacstrom, all now in Wellcome MS. 1031. This does not mean that the Bacstrom manuscripts had any relationship to George, however, as the binding does not indicate they had any commonality save being alchemical works in the collection of the architect Henry Cowell Boyes.[18]

Another contemporary individual with Rosicrucian ties was General Charles Rainsford. Although chiefly in the public eye for a long and decorated military career, Rainsford has become known since for his deep interest in alchemy, magic, and Rosicrucianism, as his manuscript collections at the British Library and the University of Pennsylvania (formerly at Alnwick Castle) attest. We do not have laboratory notes for him as we have for Bacstrom, raising the question as to whether his alchemy was more speculative than experimental.

The general was also deeply involved in a number of secret societies. Among his papers is a memorandum listing his affiliations as of 1794,

16. McLean, "Bacstrom's Rosicrucian Society"; McIntosh, *The Rosicrucians*, 87–91.

17. Glatstein, "Sigismund Bacstrom's Alchemical Manuscripts."

18. Wellcome MS. 1031.

including nearly a dozen different Masonic or quasi-Masonic bodies. He also founded a mystic brotherhood known as the "Universal Society," which boasted the miniatures painter Richard Cosway, whose work would later appear in *The Straggling Astrologer*, as a member. In his correspondence of October 1782, Rainsford mentions the discovery of Hebrew documents in Algiers regarding the Rosicrucians, and the existence of the brotherhood in that land in another form. Hills has given a possible date of 1784 for his initiation. Rainsford had ties to Bacstrom; both were acquainted with the chemist Peter Woulfe, and Rainsford made a translation of sections of Georg von Welling's *Opus mago-caballisticum et theosophicum* that Ebenezer Sibly copied and gave to Bacstrom. Nonetheless, the two different origins given for their Rosicrucian training indicate two separate lineages, assuming that both even existed.[19]

Thus, Rainsford was a self-proclaimed Rosicrucian with a wide variety of mystical and magical interests in common with George. At the time of his death on May 24, 1809, Rainsford was living at Soho Square, only a few blocks from the Grahams' lodgings on Poland Street.

The third individual to consider is Francis Barrett (1774–1818), the author—if one can use the term, given his extensive borrowings from other sources—of the influential compendium on magic, *The Magus* (1801).[20] Although Barrett's interest in ceremonial magic is well known, he also pursued alchemy. He devotes a short section in *The Magus* to the topic, and he has been credited as the author of *The Lives of the Alchemystical Philosophers* (1815), although that is a matter of dispute. Perhaps a better proof is a manuscript in the collection of the Beinecke Library, an English translation of Welling's *Opus mago-cabbalisticum*

19. Godwin, *The Theosophical Enlightenment*, 104–5, 144; Hills, "Notes on Some Masonic Personalities at the End of the Eighteenth Century"; Hills, "Notes on General Charles Rainsford (1728–1809) and His Rosicrucian Studies as Illustrated by the Rainsford Papers, Add. MSS. Nos. 23,644–23,680 in the British Museum Library"; McLean, "General Rainsford: An Alchemical and Rosicrucian Enthusiast."

20. Those with further interest should consult King, *The Flying Sorcerer*; Priddle, "More Cunning than Folk"; Sommers, *The Siblys of London: A Family on the Esoteric Fringes of Georgian England*, 259–60.

et theosophicum, with Barrett listed as the translator.[21]

Barrett bookends *The Magus* with two references to mystical societies. At the beginning, he appends the initials "F. R. C." to his name to signify his Rosicrucian membership, although he never claims the lineage. At the end appears his famous advertisement promising to teach "Natural Philosophy, Natural Magic, the Cabala, Chemistry, the Talismanic Art, Hermetic Philosophy, Astrology, Physiognomy, & c. & c." from 11 to 2 o'clock at his home at 99 Norton Street, Marylebone. Montague Summers maintained that this was the beginning of a mystical circle of students which eventually spun off a small group in Cambridge practising his rituals long after. Nonetheless, whether Barrett set up such a group is uncertain.[22]

Barrett does have one other point in his favour: an interest in ballooning. Indeed, in 1802, he made attempts to ascend from Greenwich, and did succeed—barely—in making an ascent at Swansea. In July of 1806, he sought a subscription to launch his balloon from Penzance. Given the great publicity such ascents generated, it is likely that George had at least heard of Barrett.[23]

Barrett and George also had one acquaintance in common: John Denley, the bookseller and the publisher of *The Philosophical Merlin*. According to Frederick Hockley, Barrett was on friendly terms with Denley, who lent Barrett many of the books that he made use of in writing *The Magus*. The relationship soured when Barrett failed to send Denley a copy of the book. Our evidence of these two men's ties to Denley is separated by two decades, however.[24]

Thus, we have no clear connection between George Graham and Bacstrom, Rainsford, or Barrett, or any other Rosicrucians of the time.

The other order with associations to Graham is the Mercurii. Raphael often mentions them *The Astrologer of the Nineteenth Century* as the possessors of a wide range of manuscripts on ritual magic, alchemy,

---

21. Beinecke Library, Mellon MS 140. A comparison of the Rainsford and Barrett translations has not yet been attempted.

22. Barrett, *The Magus*, bk. 2, p. 140; Summers, *Witchcraft and Black Magic*, 161–62.

23. King, *The Flying Sorcerer*, 10–17; Priddle, "More Cunning than Folk," 37.

24. Hockley, Hamill, and Gilbert, *The Rosicrucian Seer: Magical Writings of Frederick Hockley*, 19.

and geomancy, including some dating back to the twelfth and four-teenth centuries.[25]

The membership of the Mercurii remains largely unknown. Schuchard claims that the Mercurii were the result of George bring-ing Raphael into contact with his mystical friends circa 1820, although she provides no documentation for this claim. A more involved answer appears in Dicta Dimitriadis' biography of the French seer Mlle. Lenor-mand. According to this work, the seer came to London in 1822, taking up quarters on Baker Street. After spending a brief amount of time on the northwest coast of England, summoning up spirits with a "Prin-cess Olive"—most likely the impostor Olivia Serres (1772–1834)—she returned to London to bring together her new friend and ten men, known by the pen names of "Mirror," "Westminster Student," "Lilly", and "Merlinus," to create a group called the "Mercurii." She published articles in *The Straggling Astrologer,* distinguished by their large num-bers of exclamation points. After two years and a great deal of acclaim, she decided that the British were not prepared for understanding the abstruse nature of astrology, and the order disbanded when Mlle. Lenormand departed for Paris.[26]

Decker notes a number of arguments against Mlle. Lenormand's presence in London, including a letter sent from Paris in 1823 during her supposed stay in London, her intensive program of publishing in Paris at the time, and the lack of references to her stay in London in her published works. As to the latter point, Dimitriadis claims that Lenormand did indeed write a book entitled *La Sibylle à Londres,* which met with disinterest from Parisians and was returned in vast quantities. Although advertisements for such a title are noted as "forthcoming," there is no evidence that publication actually occurred, and I have been unable to locate any copy in a library. Further, a work attributed to Mlle. Lenormand, *The Oracle of Human Destiny* appeared in London in 1825. Even if we set aside Howe's speculation that Raphael was the

---

25. Raphael and Anglicus, *The Astrologer of the Nineteenth Century*, 185, 199, 215, 462, 493, 505, 509.

26. Schuchard, *Freemasonry, Secret Societies, and the Continuity of the Occult Traditions in English Language*, 486; Dimitriadis, *Mademoiselle Lenormand: La Reine de la Voyance*, 225–29.

author, the preface gives the wrong first name for the author, a difficult error to justify for anyone in a community in which she had lived for many years. Also, Raphael was advertising his membership in the Mercurii as late as 1832, an unlikely situation if the order had disbanded. It does not rule out a brief visit, or correspondence between the seer and London occultists, however.[27]

We also have the hypothesis of Marc Demarest, who has specialized in the study of the medium and occult author Emma Hardinge Britten. Demarest builds on Owen Davies' discovery that Robert Cross Smith, John Palmer, and others were members of a beneficial society for astrologers. Demarest goes on to hypothesize that this group was the same as the Mercurii, and then later metamorphosed into the Orphic Circle alluded to in the works of Britten. He has done considerable work to sketch out the connections between various individuals, but the membership of these individuals in specific societies remains elusive.[28]

Raphael's later work *The Familiar Astrologer* provides what are purportedly three sets of minutes of the Mercurii which may be dated to December 1828, 1830, and 1831 (the latter based on the publication of *Raphael's Witch*, mentioned therein). One should treat these with considerable caution as descriptions of actual events; the members often bestow unabashed praise and honours upon Raphael, and the meetings are often interrupted when a member delivers a "spontaneous" verbal advertisement for the astrologer's latest book. Nonetheless, they do give some details of the possible membership, that are at least worth examining.[29]

According to these minutes, Raphael is the secretary, and the unnamed "President" is the other titled officer. One member, the astrologer John Varley (1778–1842), is explicitly named, and Mlle. Lenormand is noted as a foreign correspondent of the group. The other attendees are indicated via initials or military titles—"Captain B.," "J. L.," "H. W.," "J. T.," "C. S.," "J. F.," "H. B.," and "C. L." Only "Lieutenant M." has a

---

27. Howe, *Raphael; or, The Royal Merlin*, 16; Decker, Depaulis, and Dummett, *A Wicked Pack of Cards: The Origins of the Occult Tarot*, 128.

28. Davies, *Witchcraft, Magic and Culture, 1736–1951*, 238–39; Demarest, "Hypotheses on the Orphic Circle"; Demarest, "A Trout in the Milk: Plotting the Orphic Circle."

29. Raphael, *The Familiar Astrologer*, 370–90, 451–56, 475–79.

likely candidate—Lieutenant Richard James Morrison (1795–1874), who would later write almanacs under the name "Zadkiel." Assuming that the other initials also accurately reflect the names of the members, these are not shared with any prominent occultist we might consider from the period, including Francis Barrett, John Denley, Frederick Hockley, Lord Bulwer-Lytton, George Graham, or even John Palmer, Raphael's successor—although the latter lists himself as a member of the society in *Raphael's Sanctuary* (1834). Also, few of these individuals bear initials in common with Marc Demarest's list of possible Orphic Circle members.[30]

Who might have been the president? For unclear reasons, Raphael tipped in a few pages into Bodleian MS. Douce 116, dated 1825. In that document, he refers to the president of the Mercurii as "H. W.," in whose horoscope the moon is sextile to Herschel. This does not correspond with the horoscope for George given in *The Astrologer of the Nineteenth Century*, but it might help future researchers narrow down the possibilities. The "President" and "H. W." are both listed in the transcript for 1830, so it is possible that the role passed to a successor, or that two members shared the same initials—or that Raphael was trying to throw others off the scent of the members' names.[31]

One clue as to George's involvement might lie in an early modern manuscript of magic. This work now rests in the collection of the Folger Shakespeare Library under the shelfmark of V.b.26. A "Charm to Protect against Thieves" in the *Astrologer* is said to come from an "Original Manuscript, dated May 8, 1577, in the possession of Mr. Graham, the Aeronaut." The charm not only appears in the manuscript, the exact date does as well. Further, a segment of the same manuscript detailing the offices of the spirits has turned up copied in a manuscript from the library of the Societas Rosicruciana in Anglia (SRIA). Frederick Hockley notes in his preface thereto that the original was "sold to Mr. George Graham, the aeronaut, for the so-called Society of the Mercurii." Nonetheless a charm on the page of the *Astrologer* after the anti-theft piece is noted as "in the Possession of the Mercurii," implying that the anti-theft charm is part of George's own library. According to Hockley, the manuscript passed through George's hands, and later ended up in

---

30. Raphael, *Raphael's Sanctuary of the Astral Art*, vii.

31. Bodleian, Douce 116, 264.

the possession of Raphael himself. [32]

Thus, if we assume Hockley is correct, it seems that George may have been a member of the Mercurii and a contributor to its library early in its history. Raphael and George first met in 1820, and that the "R. C. S." inscription on the first surviving page of the Folger manuscripts indicates his ownership began in 1822, so the Mercurii must have existed very early in their acquaintance. Indeed, it might be possible that George was a member before meeting Raphael. The aeronaut's absence from the later minutes may have signified indifference or a break with Raphael; surely Smith would have gleefully enlisted the famous aeronaut into his advertisements if possible. [33]

'The Impulse Was So Powerful': Sarah Rishman and Margaret Williams

George was married twice, although much information about these two relationships is unknown. We have only a little information about his first wife. St. James parish records relate that, on June 27, 1804, one George Graham was married to a woman of the parish, Sarah Rishman. We also have a record of a Sarah Graham, age forty-nine, who passed away of "age" and was buried on October 29, 1820. If she was indeed George's wife, then she was thirteen years his senior. Raphael describes the circumstances surrounding her passing:

> In the autumn of 1820, Mr. Graham, the aeronaut, had the misfortune to lose his first wife. While she lay ill, he had occasion to go out upon some business, leaving his wife attended by the nurse; she was in better health, apparently, than the day previous. While walking in Covent Garden Market, a thought suddenly struck him that his wife was no more; the impulse was so powerful that he could not withstand it, and

---

32. Raphael and Anglicus, *The Astrologer of the Nineteenth Century*, 504, 505; Campbell, *A Book of the Offices of Spirits*, xviii. For a published version of V.b.26, see Harms, Clark, and Peterson, *The Book of Oberon: A Sourcebook of Elizabethan Magic*.

33. It appears that there was a brief attempt to create a new organization, also referred to as the Mercurii, in Exeter in 1851. "The Members of the Mercurii."

although he hastened home with all possible speed, yet, before he could get back, he found her at the last gasp, and, before he could recover from his surprise, she pronounced a name and expired. This occurrence he related himself to a friend.[34]

A few years later, George would meet his second wife and future partner in matters aeronautical and mystical, Margaret.

Given Margaret's significance in the history of aeronautics, it is remarkable how little we know about her. In the census, she reported her birth year as 1804, in the parish of Walcot. The closest match we have is to a Margaret Williams, born to George and Hannah, baptized on December 16, 1804 in the Walcot St. Swithin parish records. I believe some data is forthcoming from the family that challenges this dating, and I look forward to seeing it. Later news stories state that one Mrs. Wilson was Margaret's sister.

Likewise, we lack any date on how she met George, or when or where their nuptials took place. None can doubt, however, that their match was a long and fruitful one, both professionally and personally. They had many children: Rebecca Frances (born April 28, 1824, 12:10 PM), Harvey George Steven James (born April 17, 1825), Rosa (born August 30, 1826, 2:38 PM; died 1890, Paddington), Alice (born June 21, 1829, 1:43 PM), George William (born July 10, 1830, 4 AM; died 1839), Lydia (b. September 30, 1832, 2:30 AM), Frances (born 1834), Jane (born 1838), Edward W. (born 1841), and Margaret (born 1844). We can be precise on many of these dates and times, thanks to a multiple-part astrological chart kept by the Grahams and passed on to the family. Most of the spotlight, however was placed on their parents and their new career as balloonists.

'The Most Gallant but Unfortunate of Aeronauts': First Flights, 1823–25

After the initial flights of the Montgolfier brothers in 1783, most balloonists had become less interested in the scientific ramifications of flight, although they invoked such inquiry as the reason for their continuing flights. Instead, ballooning in Britain was more often the work of skilled amateurs who treated the affair as a spectacle. It was not uncommon to

---

34. Raphael and Anglicus, *The Astrologer of the Nineteenth Century*, 531.

sell admission to spectators to compensate for the expenses involved, and bringing along a passenger could command large sums. Nonetheless, significant perils were involved—and not all were in the air.

As we will be spending some time with the Grahams and their ballooning career, a few caveats are in order. I will not seek to provide a comprehensive account of every ascent made by the Grahams over the course of three decades. Doing so would require not only combing regional newspapers and archives across the United Kingdom and Ireland, but also the careful sorting of intended and advertised ascents from those which occurred, as the former outnumber the latter considerably. Thus, my intent is to cover the more interesting ascents, and given the involvement of the Grahams, most of these are interesting due to the misfortunes that preceded, accompanied, and followed them. The Grahams did make many successful and uneventful ascents—but they certainly had more than their share of mishaps.

George first attempted to join the aeronauts on August 18, 1823, with the assistance of one of the sons of the famous Oxford balloonist James Sadler. George had been working on a balloon of impressive size, made of lawn—a fabric of cotton or linen—instead of the usual and lighter silk, with a coating of what appeared to be varnish. Having placed it on exhibit in the Pantheon and performed a test inflation with regular air, the men believed they were ready. They brought it to the courtyard of the White-Conduit House at Pentonville. The admission was one shilling for onlookers if outside and three and sixpence for those within to witness the inflation. Tens of thousands of people in windows, on carts, and sitting on roofs, awaited the ascent.[35]

After hours, the balloon had not filled; the *Morning Chronicle* noted that it was "not capable of retaining the subtle qualities of gas, which escaped through every part of the surface." A short rainfall saturated the fabric of the balloon, making it even less likely to ascend. Music did not assuage the restless crowd, and brickbats were thrown back and forth between the crowds on the rooftops and the ground.

After eight hours, Sadler cut the balloon loose, and he and George withdrew into the inn with the car and the money while the crowd

---

35. "Balloon Hoax and Riot"; "Mr. Graham's Balloon," August 19, 1823; "Mr. Graham's Balloon," August 20, 1823.

watched its slow, ungainly ascent. The angry spectators smashed various parts of the inn, in which would become known half-justly as the "Balloon Hoax and Riot." Such events had also occurred in previous failed balloon ascents, although whether this mitigates or exacerbates any judgment of George's behaviour is open to question. The event proved to be a fitting debut for a man who *The Times* would later label as "one of the most gallant but unfortunate of aeronauts." [36]

To gain that title, however, George needed a successful flight. With the help of Sadler and Charles Green, another prominent aeronaut, he set out to do so again. Having inflated the balloon with air at the Pantheon, George felt he was ready. He conducted it to Mr. Shewin's lumber yard on Berwick-street, Soho, on September 5. By his account, Mr. Sadler, Junior, was to accompany him, but he did not appear due to confusion as to the date of the ascent. Margaret desired to travel with him, but at the last minute, Thomas Harris, a friend of George's who was handling the ropes, leapt on board and joined him. By George's report, the two flew for fifty-five minutes, flying one and three-quarters miles, rising to an altitude of two and a quarter miles. [37]

On September 12, George made another attempt at the White Conduit-House with Sadler or a man named Hall. The balloon inflated slowly, but eventually it seemed to have enough gas to bear the two men. As they began to rise, however, a trailing rope caught on a pole. The crowd's efforts to free it snapped the pole, causing the upper part to slash a yard-long gash in the fabric. The men escaped the balloon, but by that time it had partially righted itself. George leapt back in, allowing him to ascend for the brief space of five minutes before descending in a gravel pit. Nonetheless, a modest success was still a success, and George Graham's aeronautical career began. [38]

George seems to have switched balloons for subsequent flights, likely to a more conventional one of silk. A year later, a pastry cook named Smithies attempted to take up the "lawn balloon" to generate

36. "Ascent and Descent of Mr. Graham's Balloon."

37. "Mr. Graham and His Balloon," September 6, 1823; Graham, "Mr. Graham's Account of His Ascent."

38. "Ascent and Descent of Mr. Graham's Balloon"; "Mr. Graham's Balloon," September 13, 1823.

support for Greek independence, but it failed to inflate. A representative of the Imperial Gas Company said that his business would not only provide no more gas, but also they would have refused service if they known the "lawn balloon" was being used. Smithies then lent it to a Mr. Courtney, who attempted to ascend in it at the Green Man pub on the Kent-road. This failed as well, and the pub's owner kept the balloon as surety. Smithies went to court to recover it, but he was unsuccessful. Apparently, this leaky contrivance would make its way back to the Grahams, as shall be seen later.[39]

George's next ascent occurred on September 24 in Oxford. His craft knocked slates off houses and set down in the middle of town, where his passenger disembarked. Freed of the weight, the balloon shot up into a cloud, drenching the aeronaut. The grapnel he employed to slow the balloon on descent tipped over the car, causing him to be dragged for nearly a mile, scattering his telescope, book, and other contents of his pockets.

Sadly, George's ballooning enthusiasm had infected Thomas Harris, his passenger on his very first flight. Harris had served as a mate on a ship to the East Indies, and after returning to London, lived on Wells Street and took up the trade of cabinet-maker. He met George when the aeronaut sought someone to perform work on an unspecified aspect of the balloon. His intent was to fly in a balloon himself, without the assistance of George.[40]

Indeed, George offered little help to Harris in his new avocation. Raphael notes cryptically that "although they were at this time [of their flight] excellent friends, yet soon afterwards (partly through female intrigue), they became decidedly hostile to each other." As Harris' flight grew closer, George published a handbill to dissociate himself from the event. He insisted that he was "not in the least connected with the Machine" in which his former friend would ascend. Further, Harris had had no hand in the creation of George's balloon or in the expenses of the flight. In fact, the man's ascension with him was at the last-minute instigation of a friend, and "Mr. Harris merely accompained

39. "The Grecian Balloon"; "Court of Common Pleas, Friday, May 27: Smithies v. England."

40. "Fatal Aeronautic Excursion."

[sic] him as *Ballast* and to *Balance the Car*, & c." It is not clear as to whether George was suggesting he might have thrown Harris from the car if necessary.[41]

Yet this lack of support did not deter Harris. If we are to believe Raphael, neither did a visit to the noted astrologer, during which Smith told Harris his chart foretold disaster unless the flight was delayed until after the first week in June. The celebrated balloonist James Sadler attempted a more mundane intervention, with no luck. Harris was set in his course.[42]

On May 25, 1824, resplendent in a blue naval uniform embroidered with gold, Harris prepared for his ascent from the Eagle Tavern Gardens. Sophia Stocks from the crowd volunteered to accompany him. The balloon flight was smooth, until Harris began his descent. It is unclear as to what happened, but it seems that Harris pulled the wrong rope and opened a large valve at the top, releasing most of the gas. The balloon came down precipitously in Beddington, where it struck a tree, killing Harris and seriously injuring Miss Stocks.

George was on hand quite soon afterward, interviewing the lady the following day while accompanied by a member of Harris' ballooning committee. He was called as an expert witness at the inquest, who agreed with the suggested cause of accident. To give some idea of the tenor of the time, the tree which Harris had struck was nearly torn to pieces for souvenirs.[43]

In what was likely an effort to reassure the public, Mr. Graham took another flight from Pentonville on June 2. The *Westmoreland Gazette* provides an interesting description of the balloon used for the flight:

> ...it was formed of alternate stripes, eighteen inches wide, of pink and white silk, and round the middle was inserted a complete circle of cerulean blue... the car which was placed near the balloon was covered and lined with crimson and silk velvet, bearing on one side the arms of England, on the reverse

41. Raphael and Anglicus, *The Astrologer of the Nineteenth Century*, 360; Graham, "To the Public."

42. Raphael and Anglicus, *The Astrologer of the Nineteenth Century*, 438; Davies, *King of All Balloons: The Adventurous Life of James Sadler, the First English Aeronaut*, 235.

43. "Account of Mr. Harris' Ascent, and the Fatal Result That Attended It."

those of the City of London, surrounded by gold rays, and supported by doves with the olive branch on each side, encircled also with rays of gold. At the bottom was represented Aeolus calming the fury of the winds, with pedigree ad infinitum—

"Ventorum at Libs, Notus, Auster.

At either end were the figures Fame and Victory in carved work, richly gilt and burnished, the whole entwined by the rose, the shamrock, and the thistle, and decorated with gold fringe and tassels. Over the car, surrounded with gilt pillars, was a canopy, round which were represented the Zodiacal signs, hung with a festooned drapery of rich crimson satin, and bordered with gold fringe tassels. Previous to the fastening of the car to the balloon, the crimson drapery that covered it was withdrawn, and on the sides was painted the Council of the Gods, with Jupiter in his judgment seat, pointing upwards.... [44]

August as his conveyance was, George was to inflict one more indignity upon Harris: his balloon passed, by ill wind or poor taste, directly over the man's funeral procession. Mr. Graham would later write the *Times* to insist that the date of his flight had nothing to do with Harris.[45]

That October, George embarked on a series of ascents at Bath, first from Sydney Gardens and later, due to a clogged pipe at that location, the yard of the Bath Gas Light & Coke Company. Most of these were underwhelming, facing the usual problems of insufficient gas and lift. This series of launches did see Margaret ascend to the heavens for the first time, accompanying George on October 28. The Grahams would return to Bath and Bristol repeatedly over the coming years.[46]

On October 13, George set out after another ascent at Hammersmith, but he ran into an unusual spot of trouble. By this time, he had apparently set up a Managing Committee for his ventures that included a

44. "Graham's Balloon."

45. "Death of Mr. Harris the Aeronaut"; "The Inquest"; "Mr. Harris and Miss Stokes."

46. Penny, *Up, Up and Away!: An Account of Ballooning in and around Bristol and Bath 1784–1999*, 10–15.

Mr. Fargues, a tradesman from St. James' parish, and a picture-dealer from Duke's-court, St. Martin's lane, named John Adams. Adams would be a later companion and advisor, and occasional passenger, for many subsequent flights.

The three men, riding in a cart, were riding the toll road from Hammersmith to Hyde-park-corner. When arriving at the Kensington gate, the toll-collector, George Noble, grabbed the horse and pushed the cart back. When Mr. Adams attempted to show him the ticket, Noble attacked him and then rounded on Fargues as he attempted to read the man's name above the door. George held the reins, and eventually all three men drove into town, where they reported Noble to the authorities. Noble was arrested and brought before the magistrate, who granted him bail of a significant amount. George attempted an ascent at Hammersmith twice, on the 18th and 19th, and he put off the crowd by promising a third, which seems not to have occurred.[47]

The Grahams' lives and the Pantheon would once again intersect on October 28, 1824, when one Nicholas Cundy sought to renew his license to the building. Cundy was already in serious trouble; he had had considerable difficulty obtaining a license for entertainment from the Lord Chamberlain. His subsequent efforts to circumvent these from 1811–13 through creative definitions—such as labelling theatre with music as "musical entertainment"—had earned him the disapproval of the judiciary and landed him in jail for debt for four years. The solicitor's clerk for the shareholders responded that the property was not in Cundy's hands, but rather in those of George Graham. George was not only using the Pantheon as a site for storing his balloon; the stockholders were paying him to keep an eye on the building. Cundy objected that George had been absent for three weeks, but he was forced to withdraw the suit.[48]

George made several flights in late 1824 and early 1825. One ascent from Pentonville on May 11, 1825 was most notable for his choice of passengers. A "gentleman, who stated himself to be the Editor of a forthcoming work on Astrology, tendered a sum of money (we understood

47. "Police: Marlborough-Street."

48. London City Council, *Survey of London*, vol. 31–32, pts. 2, 268–83; "Middlesex Sessions, Thursday, Oct. 28."

30£)" as a passenger. The most likely candidate is Raphael, who would release *The Astrologer of the Nineteenth Century* that year. Given what we know of him, it seems unlikely that he had thirty pounds to spend on a balloon ride, but the newspaper account expresses uncertainty as to whether the money was produced. At any rate, a Captain Currie, of the Third Dragoon Guards, outbid him with 40£. The captain had a quiet flight, with no mishap other than the balloon brushing some trees on lift-off, and he and George would be associates for some time to come.[49]

George Graham's ascents were not just displays of aeronautic prowess, or lack thereof; they also brought a great deal of entertainment to onlookers. Crowds for some of the ascents ranged up to tens of thousands of spectators, and other entertainments were in evidence. One ascent at Mile End in May of 1825 featured not only his balloon, but also live music, a Punch and Judy show, the "tallest man and smallest woman in England," and an abortive fireworks display. [50]

When preparing for an ascent at Chelmsford in July, George agreed to take on board Captain James Gape. Gape was a daring man and officer in the Royal Scots, who had ridden off the field at Waterloo with two musket balls embedded in his saddle. As might be expected, George soon found that the balloon was only capable of carrying one passenger. The Captain offered to ascend alone, with no experience in such matters, and had to be persuaded to relinquish his seat. As George was ascending, Gape suddenly leapt into the basket, knocking the basket sideways and pulling the car down toward the ground. The balloon caught on a chimney at the House of Correction, dangling the Captain over a *cheval de frise*—that is, a portable barrier covered with long spikes intended to stop cavalry. He made it safely to the roof, and George was able to descend without incident.[51]

---

49. "Balloon Ascent," May 12, 1825.

50. "Balloon Ascent," November 18, 1825.

51. Wood, *Those Terrible Grey Horses: An Illustrated History of the Royal Scots Dragoon Guards.*, 61; "[Untitled]," August 21, 1825. Although the timeline is unclear, George might have failed an ascent here the following month. Charles Green had to take out a local advertisement four years later stating that not he, but Mr. Graham, was responsible for this disappointment, so that his own ascent would not lose customers. "Balloon and Parachute."

'To Be So Engaged SPEAKS VOLUMES!': The Controversy in Norwich, 1825

Once the balloon had been repaired again, the Grahams travelled to Norwich. There they offered a ride to John Harvey, the seventy-year-old High Sheriff of Norfolk, from Richmond Hill Gardens. It will be no surprise that, even after repeated and exhaustive preparations, George declared the balloon was unfit to carry more than one person. As an additional hurdle, he was not using coal gas as he had in the past. Instead, he was attempting to create hydrogen gas via iron and sulfuric acid on site, a process that he had never attempted before. The Sheriff declared his desire to fly alone, although the crowd called upon him to be accompanied by someone. Oddly enough, the person who took them up on it was Captain Gape, who briefly occupied the basket with Harvey. Eventually both were coaxed out, and George made a brief trip aloft.

The balloon remained slightly inflated thereafter, which George saw as a hopeful sign that he could inflate it fully and take up the High Sheriff the next day. Nonetheless, the local paper reported that someone had opened the aperture to allow most of the gas to escape. The following week another suspected sabotage of the valve occurred, but the balloon was eventually inflated. It was questionable whether it would ascend with two people, but after George was replaced with Captain Gape, and Gape with Margaret, the High Sheriff finally ascended for an uneventful flight.[52]

Ordinarily, this would have been the end of the matter, but the failed ascents had become a scandal locally, especially due to the statement of William Stark, a local chemist and dyer who John Harvey, being suspicious of the Grahams, had engaged to watch the inflation. Stark in turned called upon his friend Shakespear Bell, visiting from London, to assist him. As Stark himself would later be elected as a Fellow of both the Geological Society of London and the Chemical Society, his expertise provides useful insight into the Grahams' process. Stark's statement condemning the Grahams was published in the September

---

52. "Norwich, September 7"; "Colonel Harvey's Flight"; "The Balloon."

10 issue of the *Norfolk Chronicle and Norwich Gazette*. [53]

Margaret was still in town with her children. As it turned out, little Harvey Graham was baptized at St. Peter, Mancroft three days after Stark's comments appeared. Nonetheless, Margaret took little time in making her only foray into publishing with the *Statement of the Particulars Relative to Mr. Graham's Balloon, Explaining the Cause of Disappointment to the Public, the Heavy Loss He Sustained from It, and an Account of the Receipts and Expenditure*. The *Statement* was printed at the Norwich printer W. Booth, who had also released an account of her flight with the Sheriff. The book was placed on sale there and at an establishment of Mr. White, a cobbler. Her introduction, dated September 29, mentions that Margaret was "detained in [Norwich] much longer than she had anticipated" due to Mr. Graham being "in confinement in a strange city." The failure to ascend, she asserted, was due entirely to the failure of the chemists, who had used up the raw materials quickly, failed to order more materials in advance even after being encouraged to do so, and allowed a great deal of gas to escape through improper use of the apparatus.[54]

Not to be outdone, Stark fired back with a pamphlet of his own vindicating his previous statements. As this author is not a chemist, I cannot speak as to the validity of the claims made in either pamphlet. Stark nonetheless acquits himself as a knowledgeable individual, precise in his statements and measurements. While both claim that others present will validate their accounts of events, Stark acquired statements from Bell and one of the other individuals involved in order to corroborate his account. All in all, he makes a much more convincing case.[55]

If we take Stark's account as valid, the ascent preparations were a fiasco. George had little idea of how the chemical reaction to create hydrogen worked, inadequate equipment, and insufficient ingredients

53. The Chemical Society of London, "Proceedings of the Meeting of the Chemical Society," 436. The issue in question has not been included in digital archives at this time.

54. Graham, *Statement of the Particulars Relative to Mr. Graham's Balloon, Explaining the Cause of Disappointment to the Public, the Heavy Loss He Sustained from It, and an Account of the Receipts and Expenditure*, [iii], 9.

55. Stark, *A Letter to John Harvey Containing an Examination of a Pamphlet Entitled, "Mrs. Graham's Statement of Facts" Relative to the Balloon.*

to bring about the inflation. Specifically, George underestimated the amount of iron available and did not sufficiently clean what he had in order to bring about the reaction. The Grahams were forced to cast about quickly for more supplies of both ingredients, having them conveyed with some expense. Further, many observers saw signs that the gas was escaping the balloon, even as George was purportedly filling it.

Stark also gives his own interpretation of the events surrounding the ascent. Stark claimed to have witnessed "secret signs" among the Grahams and their people, suggesting that they were prolonging the process for the sake of gaining more revenue from onlookers. He witnessed a Mr. Bramble, in their employ, crawling beneath the balloon out of sight, followed rapidly by a diminishing of the balloon's inflation. Further, George had decided on the day of the ascent that he would not be able to send the balloon up. He hoped that the soldiers would protect him and the balloon as it was borne away, but Bell told him this would be insufficient for his safety, and George was convinced to make the ascent. Even more scandalously, a man sought to find Mr. Bell the day before the ascent, when he discovered the following:

> ...he was directed to a room upstairs, which was a *bed room*, in this he did not find Mr. Bell, but he found Mr. Graham *sparring* with another man; two or three other persons were in this room, and amongst them a LADY *veiled*, viewing this sparring exhibition. Now, Sir, for this man at such a moment to be so engaged SPEAKS VOLUMES![56]

Aside from Graham's clear distraction from ballooning, this incident might not speak volumes to us in the same way that it did at the time. Although illegal, pugilism had seen a modest amount of support from nobility. By the 1820s, however, the clientele was almost entirely from the working classes. Prize fights were raucous affairs that brought with them unruly crowds and heavy betting. It was not unknown for bets not to be honoured or the prize money stolen, and it was much more common for a contestant to be paid for throwing a match than for it to be fought fairly. Thus, Stark was associating George with an activity

---

56. Stark, 24.

that would be seen as low-class, dangerous, and of questionable probity.[57]

After the launch, the Grahams submitted a vastly inflated bill for services and followed this up with an "Appeal to the Public," apparently an effort to take up a subscription to cover the costs of the ascent. Nonetheless, George ended up owing fifty pounds, and he had gone to debtor's prison briefly while it was worked out. At the time the pamphlet was published, some tradesmen had yet to be paid.[58]

Splashdowns, Landlords, and Court Cases: The Grahams, 1825–35

Somehow the Grahams managed to extricate themselves from Norfolk in time for their next flight. They took to the air at Plymouth on November 14, 1825, inflating their device at the Stonehouse Marketplace after two abortive attempts. An employee at the dock-yards, one Mr. Grills, was intended as a passenger, but did not accompany them. The ascent went smoothly, taking them a mile into the air very quickly. They decided to put down on the Mewstone, an island off the coast. Their aim was off, and the balloon landed in the water, the car being below the surface. The balloon pulled them along for twenty minutes, as the Grahams clung to the gondola. A passing boat eventually caught up with them and dragged them out of the water. The balloon's ropes became entangled with the mast, leading two sailors to be seriously injured as they cut it loose. The near-dead Grahams were conducted back to land, and sailors later recovered the balloon. Margaret Graham later claimed that the two had known that they would be blown out to the ocean on this flight, equipping themselves with life preservers for the inevitable crash. According to the *Times*, George would later claim that his balloon had flown to the Mediterranean, where it was picked up and brought to Gibraltar, from whence he could not retrieve it.[59]

January of 1826 found George, along with a scientific partner named

---

57. Brailsford, *Bareknuckles: A Social History of Prize-Fighting*, 79–91.

58. Stark, *A Letter to John Harvey Containing an Examination of a Pamphlet Entitled, "Mrs. Graham's Statement of Facts" Relative to the Balloon*, 16.

59. "Balloon Ascent," November 18, 1825; "The Late Balloon Ascent"; "Balloon Ascent from Glocester."

Webb, in court. The two men had desired to fly together, but the balloon needed repairs, which were entrusted to a tailor named Heely. Heely completed the repairs, but George and Webb did not pay. The judge ruled for Heely and suggested that the parties work out the matter. Webb does not appear to have made an ascent with George the previous year, or at any time going forward.[60]

Margaret Graham's first solo ascent occurred on June 28, 1826, at the Grahams' favourite launching grounds at the White Conduit-House, Pentonville. Although she intended to ascend with Miss Stocks, Harris' companion on his ill-fated trip, Margaret found the balloon insufficiently inflated for two. In fact, it was barely inflated enough for one, as Mrs. Graham's releasing of the ballast caused her merely to skim over second-story rooftops. At one point, the balloon became caught on a house's trim, and she had to push off with her foot. The balloon descended at Newington Green.[61]

While Mrs. Graham enjoyed the hospitality of a local family, Mr. Graham and Mr. Adams, his partner at the time, attempted to pack the balloon into their gig. A crowd accosted the two men, who demanded to be given beer so that they would not be able to injure the balloon. After they ran through the sixteen gallons obtained from the Green Man Public House, they redoubled their efforts at sabotage. The balloon was taken to the Green Man, and later to a tenant of the proprietor, Mr. King. King later presented George with a bill of 20£ for the beer and damage to his property. The aeronaut refused to pay and managed to retrieve his balloon after repeated efforts, which King claimed had contributed to a riot at his premises. The magistrate at Worship Street demurred from making a judgment for either side.[62]

George made some ascents at Bristol later in 1826, but his last attempted flight at Birmingham, on October 16, led to his arrest. Two days before, his balloon had been seized by creditors and held at the Shakespeare Rooms. The morning of the flight, George led a group of

---

60. "Law Intelligence, Court of King's Bench—Monday, Jan. 9: Air Balloons—Heely v. Graham and Webb." A Mr. Webbe made some ascents with George in 1823, but it is unknown whether this is the same individual.

61. "[Untitled]," June 29, 1826.

62. "Mrs. Graham, the Aeronaut"; "Police: Worship Street."

men and seized it back, supposedly by force. While he was preparing the balloon for flight in front of the crowd, however, an officer of the sheriff appeared, delivering a writ of execution and seizing his goods, including the balloon. The *Birmingham Chronicle* reported later that the overzealous officer also seized the money the crowd had paid for the ascent, which was still technically the possession of the onlookers until the ascent had occurred. All of this was unclear to the crowd, and different theories were in circulation as to what had actually happened. George himself claimed that bettors, having placed large bets against the success of the ascent, had held down the balloon until the gas escaped.[63]

Retrieving the balloon would be more difficult this time. The Grahams had partnered in their aeronautical enterprise with one Charles Bailey. Bailey seems to have been strictly engaged from the business end; when he made an ascension of his own with Charles Green on June 7, 1827, he spent the flight curled up at the bottom of the basket. At any rate, he had gone 361£ in debt to bring about his partnership with George, and George had recently sold Bailey his share in the balloon. The partners were further in debt to three men—Mr. Brooker, William Maton, and Morton Grafton.

After the sheriff confiscated the balloon, it came into the possession of Brooker, who offered to sell it to Maton and Grafton, while holding Bailey not responsible for his portion of the debt. As Bailey was still part owner of the balloon, the transaction came into difficulties. Mr. Brougham, representing the defence, pointed out that only four aeronauts were in England at this time, so the market for the balloon was in any case quite limited.

Margaret Graham appeared at the trial, where she testified as to the balloon's worth. George had spent over 260£ and four or five weeks, with the help of many people, in making it. This balloon, she claimed, had only ascended twice. She had negotiated with Grafton at the offices of a Birmingham lawyer for its return. In the end, the judge ruled that the plaintiffs were to be granted 50£ in damages. The balloon itself was sold off and came into the possession of an H. Green (not to be

---

63. "Court of King's-Bench, Westminster, June 8, Maton and Grafton v. Brooker"; "Mr. Graham and His Balloon," October 21, 1826; "Questions Continued: X."

confused with the famous balloonist Charles Green), who repeatedly attempted a flight at Devizes three years later. The balloon failed to inflate, and the crowd cut it to ribbons.[64]

It is not clear how George once again left incarceration, but he does not seem to have caught up with his ballooning until September of 1828, when he made an ascent in Southampton. Later that year, Margaret went up twice at Chichester. Following the second descent, the balloon was paraded through the town. At one point, it caught on a pinnacle of the monument known as the Chichester Cross, which may still be seen in that town. Part of the Cross toppled into the crowd, but no one was hurt.[65]

Margaret's last flight for the year occurred at Northampton. As might be expected, her intent to bring along a Mr. Pickering as a passenger was defeated due to underinflation. Indeed, soon after take-off, observers noted a foot-long rent in the balloon's side. This and other tears in the fabric prevented the balloon from ascending over the surrounding buildings. The balloon became entangled with a chimney, sending mortar and bricks showering down on Margaret, who sought to escape through the attic window of one Widow Ager. Freed of her weight, the balloon rebounded into the air. The gondola fell into the river Nene, and the balloon itself eventually landed near Tausor. According to the *Northampton Mercury*, the locals initially believed the balloon was a sign of the Second Coming. This was followed by a discussion of shooting it, an attempt to scale it by ladder, and the cutting open of the fabric in hope of releasing any people trapped inside. A local farmer eventually dissuaded his fellows from completely destroying the balloon and returned it to town free of charge. Even with this act of generosity, the flight cost four times the contributions it brought in.[66]

Following Northampton, the Grahams went on a hiatus from ballooning. We have one hint of a possible second career for George during this time. In February of 1831, George placed an advertisement in the

---

64. "Court of King's-Bench, Westminster, June 8, Maton and Grafton v. Brooker"; "[Untitled]," August 1, 1829.

65. "A Ghost"; "Balloon Ascent," October 22, 1828.

66. "Most Perilous Balloon Ascent"; "The End of the World"; "Balloon Ascent," November 7, 1828.

Morning Advertiser for the sale of a cosmorama, an apparatus that allowed viewers in a dark room to view impressive pictures through a lens. Was this merely a quick transaction, or did this signify the end of a side venture for the Grahams?[67]

Ballooning seemed out of the question. A John Algar, auctioneer at 9 King-street, Holborn, announced the sale "by the order of the Sheriff of Middlesex" the balloon, car, and other paraphernalia previously in possession of the Grahams. The first notice appeared in the *Morning Advertiser* in May 13, 1831, and in three subsequent notices, the last being on June 9. By the 16th, he seems to have been unburdened of it, as it ceases to appear in his advertisements.[68]

Somehow, the Grahams managed to buy back or otherwise obtain a balloon, leading to an exalted triumph on September 3, 1831. Their ascent at Windsor was delayed twice, and the third time, the message came that the king himself wished to view the balloon's flight. They were able to carry off this ascent successfully, save that the balloon was once again underinflated, so George had to ascend without Margaret.[69]

The Grahams' balloon becomes the topic of an 1833 suit at the Court of Common Pleas. According to the plaintiff's testimony, the Grahams owed a silk merchant, Mr. Bennington, 68£ for materials for a balloon. The bill not being paid, Bennington sought out a writ of execution against them, and a sheriff's officer, Mr. Davies, was charged with confiscating the balloon. Supposedly Davies sent an agent, who had been drinking at a public house, to the Pantheon to retrieve the property. Once arrived, Mrs. Graham accosted the agent. It was purported that she got him even more drunk and possibly offered him a bribe. No matter what occurred, the silk balloon he had been sent to retrieve was switched for one of cotton—no doubt the infamous "lawn balloon" that began George's career. This balloon arrived at the auction house, which sold it for the princely sum of 4£ before anyone caught the switch. Bennington took the sheriff to court over the matter. The *Times* reporters were going to refrain from reporting it until an

---

67. "Cosmorama to Be Sold."

68. "Sales by Auction," May 13, 1831; "Sales by Auction," May 16, 1831; "Sales by Auction," May 27, 1831; "Sale by Auction"; "Sales by Auction," June 16, 1831.

69. "Balloon Ascent," September 5, 1831.

unnamed party offered them a bribe to keep it quiet.[70]

Around this time, the Grahams also had to deal with a more mundane problem: the reacquisition of a landlord. George and Margaret did not own the house at 41 Poland Street, but the ownership of the property had apparently been forgotten for some time, as the tenants had not been asked to pay rent for nearly three decades. When some work was being done on the Pantheon, a joint wall with the Grahams' property was torn down. The aeronaut visited an attorney, a Mr. Bromley, who suggested that George should become his tenant, so that some redress for the wall could be obtained. This led to Bromley obtaining a 90-year ground rent on that part of the property, with a charge of 30£ for five years and 60£ for the following ten.[71]

Although Bromley asked for no rent for the first year, he later put up the rent for sale, transferring it on February 27, 1835 for 365£. The rent came into the hands of a Strand grocer named Moore. George seems to have been put out at this situation, later saying, "I was not so *wide awake* as I was afterwards" when he made the decision to pay the rent. Nonetheless, it does suggest one of the economic factors that allowed the Grahams to engage in their aeronautical adventures was their previous lack of rent for many years.[72]

It was about this time that Margaret became the primary aeronaut in the family. George was still very active in the commercial aspects of the business, and he made a point to follow Margaret in a post chaise, so the balloon could be retrieved. It is likely, however, that Margaret's sex made her a greater draw than George, and he had other interests that were coming to the fore, including astrology, alchemy, and ritual magic.

---

70. "Court of Common Pleas, February 6."

71. Great Britain Central Criminal Court and Buckler, *Central Criminal Court. Minutes of Evidence*, vol. 12, vol. 12:713; "Diary of Public Sales, Feb. 27."

72. Great Britain Central Criminal Court and Buckler, *Central Criminal Court. Minutes of Evidence*, 12:713; "Diary of Public Sales, Feb. 27."

'Gold Will Be as Plentiful to Me as Dirt': George Graham and Astrology, Alchemy, and Ritual Magic

Their interest in ballooning did not mean that the Grahams were idle in occult matters. Indeed, it was for a time a key part of their family's livelihood. In Harry Price's "Wonderful Magical Scrapbook" at the University College of London's Senate House Library, we find an advertising card bearing the Grahams' address, 41 Poland Street. Based upon its reference to Raphael, it dates after his death in 1832, but the activities described within may not be confined to this period. It reads as follows:

ASTROLOGY.

Ladies are respectfully informed that a variety of Bonnet Shapes and Stiff Sleeves, are kept in sale at 41, Poland St. Oxford St. N. B. Every purchaser of the above articles will receive answers to questions on the most important events of life gratuitously, on Astrological principles, by a Pupil of the late RAPHAEL, author of the Prophetic Messenger, & c. Private door.

It is likely that this elaborate pretence for providing astrological consultations was an attempt to circumvent the 1824 Vagrancy Act, under which astrologers were often prosecuted. It might also be that the appeal to women was intended to keep male customers—and possible policemen—away. A note next to this item in the Wonderful Magical Scrapbook indicates that five hundred copies of this card were printed when the two went into business.

We might ask whether the unnamed "Pupil" noted was George or Margaret. After all, George had ties to Raphael, but he seems to have been established in his knowledge of occultism by the time he met Smith. As a point in Margaret's favour, a "Ladies' Astrological Society" at one time held meetings at 38 Poland Street, not three doors from the Grahams—and possibly another address at which the two conducted business, as we find in the curious matter of the Astronomical and Astrological Society of Great Britain. I say "curious" because we have only the outline of a controversy, most of the details of which are still obscure. Our starting point is a notice in the *Athenaeum* for May 10, 1834:

ASTRONOMICAL AND ASTROLOGICAL
SOCIETY of GREAT BRITAIN
Instituted April 10, 1834, at Noon.

Professors and Amateurs of Celestial Science may receive every information concerning the objects of this Society, by addressing a letter (post paid) to the Secretary, at Mr. Denby's, Bookseller, 24, Brydges-street, Covent-garden.[73]

A letter in the Scrapbook, sent to a Mr. Northwood by Society secretary Thomas Gale, gives more information about the early membership of the group. The first meeting saw twelve members join the group. One J. Day was named president, an R. R. Manger was vice president, and J. Johnson was the treasurer. George was not listed as a member, but then again, neither was Thomas Oxley.

We know relatively little about the personal life of Oxley (1789–1851). He lived in London, Liverpool, and possibly the United States, and wrote many books on astrology and other topics. He is perhaps best known for his strong advocacy of circular astrological charts, which eventually replaced the rectangular ones which astrologers had used for centuries. Oxley soon took over the leadership of the Society. We have two pieces of stationery for the "Astronomical and Astrological Society of the United Kingdom" available to us, one of which has the original officers crossed out and Oxley's name written in as the group's president, with J. Johnson as the vice president and a Randolph Oxley as secretary.[74]

73. "[Advertisement for the Astronomical and Astrological Society of Great Britain]." 'Denby's' is likely 'Denley's.'

74. Gansten, "Placidean Teachings in Early Nineteenth-Century Britain: John Worsdale and Thomas Oxley"; Wonderful Magical Scrapbook. In 1841, Randolph Oxley would take possession of a Frederick Hockley manuscript on crystallomancy, giving himself the title of "G[rand] M[aster] of the Rosy Cross." Godwin, *The Theosophical Enlightenment*, 130.

George and Oxley seemed to get along for at least a time. When Oxley proposed to give "A Course of Two Interesting Lectures on Astronomy and Astrology" on June 24 and 26, 1834, it was not only at 59 Poland Street, just down the road from the Grahams, but a "Graham" was selling tickets to the event from an address at 38 Poland Street. The day before the first lecture, however, relations had soured—possibly, as we shall see, for some incident involving Margaret—as the Scrapbook provides the following document:

A List
of those members of the
Astronomical and Astrological Society of
Great Britain

who agreed to meet on Monday June 23rd 1834 at 6 PM precisely: to take into their consideration the Best and most effectual means of remedying the present evils which have crept into the Society and menace it with inevitable ruin unless speedily removed.

Also for the purpose of electing another President,
Mr. Oxley having greatly aggressed upon the Laws
of this Most Noble and Usefull Society

The only signatory on the list is George Graham. His assessment of the Society's prospects seems accurate, as we hear nothing of it after this date, less than three months after its foundation.

George's alchemical interests also continued at this time. Two alchemical manuscripts that include sections written by the Grahams are kept in the collection of the Wellcome Institute. Unfortunately, both are quite fragile and difficult to view for reasons of preservation, but the glimpses I have had of them confirm that both Grahams were quite engaged with this topic. We will discuss Margaret's interests later, as they seem to post-date those of George by many years.

The first of these works, Wellcome MS. 1031, comes to us from the collection of Henry Cowell Boyes, passed on to Edward Matthey in 1894. It is bound with three treatises on alchemy written by Sigismund Bacstrom, although it cannot be determined whether these were in George's possession or bound with his notebook later. George's section begins with a couple of newspaper clippings, along with large letters proclaiming the work to be "'Alchemy. George Graham. Student. Pantheon, Oxford Street. 1824. London.' The catalogue notes that the watermarks on the paper date the work to 1825 and 1827, and a newspaper story regarding a quicksilver mine from November 25, 1838 appears within, so the notebook served him for over a decade.

After a few pages, we arrive at a page with the longer title "Alchemy: A select collection of Testimonies respecting the Doctrines and practice of the Ancient Alchemists Extracted from their Writings." Several chapters follow, each filled with extracts from different alchemical authors on different topics—"Of Sulphur," "Of Our Mercury," "Of the Secret Fire," "Of Rebus," "Of the Three Principles," "Of the Furnace and Glass," and "Of the Work." Some other treatises break up the order, but these seem to have been inserted later by Margaret.

What we lack is any notebook detailing the Grahams' own experimental work in alchemy, but he certainly attempted to undertake it. On March 5, 1833, George wrote the occult bookseller John Denley with the following request:

> Sir, having perfectly satisfied myself that I am in full possession of the whole secret of the transmutation of mettals, I shall take it as a favor if among your Connexion you could introduce me to a to a [sic] gentleman studying the science, one who would join me in perfecting the matter. Any gentleman at all conversant with the various authors on the subject, I am sure, would in a short conversation be satisfied that my assertions be well founded. I have only to add that I would not join one who had not studied the Art a little.[75]

In 1839, George was still pursuing his alchemical speculations, as a letter to Alexander Byrne outlines. George seeks from the man an investment in his alchemical purposes—fifty pounds to build an appropriate laboratory and two pounds a week for his salary, along with half the cost of the necessary materials. In return, he promises to complete the "great Magistry" after nine months, or a simpler and less certain process after six weeks. If he or one of Byrne's friends is unwilling to

---

75. Wonderful Magical Scrapbook.

put up the funds, George was willing to "dispose of the lease" to his house in order to raise them. The response received to this letter, if any, is unknown; Byrne perished in the wreck of the SS President in 1841, so any collaboration would have been short-lived at best.[76]

How aware was the public of the Grahams' interests? In addition to them being mentioned in Raphael's nativity of George in *The Astrologer of the Nineteenth Century*, we have a satirical piece in the Wonderful Magic Scrapbook—sadly, with no date or place of publication. We can date the piece to late 1836, or the following year, based upon the references to the Duke of Brunswick as a passenger of the Grahams.

The article's premise is that London astrologers have been put out by one 'Paul Pry'[77] who is exposing their activities. These worthies engage in all sorts of antics, considering an assault upon Pry that their incompetence and arrogance will prevent them from carrying out. The Grahams enter the imaginary meeting, as Mrs. Graham proposes to get Palmer's underwear washed:

> She was going on at some length, when he husband requested her to be seated. Mr. GRAHAM, the Aeronaut, Alapemist [Alchemist?], and Astrologer, began with, "My dear children, I admire the preceding remarks, and I know how the students are oppressed, but don't mind, I have discovered the powder of projection, all I want now is one simple ingredient, when I discover the stone, gold will be as plentiful to me as dirt; I will then take you all by the hand, Mrs. Graham and all the young Grahams shall have a new balloon a-piece, Capt. Curry, and Duke of Brunswick, and every astrologer shall have a balloon of his own, I will make you all independent, you shall change the name of astrologers to ballooners out of gratitude to me…"

Margaret proposes that some of this gold be used to bribe Pry, but Palmer tells her that the journal's editors are too principled to accept it. After more discussion, George falls asleep, and Oxley takes the opportunity to mesmerise Margaret. George objects when he awakes,

---

76. *Ibid.*

77. 'Paul Pry' was a character in a famous 1825 play who uncovered all manner of misdeeds with his inquisitive ways.

and this leads to a general melee between his and Oxley's supporters. The police arrive, and all the participants flee.[78]

Another area in which George Graham had an interest was ritual magic. Summoning spirits via incantations, magic circles, and consecrated tools might seem more fitted for centuries before, but it was nonetheless a key part of the London occult scene, especially the circles in which the Grahams travelled. Barrett's *The Magus* revived readers' interest in such arts, Raphael accompanied his astrological predictions and essays with magical procedures and stories, and the bookseller John Denley sold both printed works and custom manuscripts on the topic. George himself displayed an interest in the subject, although we do not know how great it was or if it ever inspired any practice on his part.

One important manuscript of sixteenth-century ritual magic passed through George's hands. This work, now owned by the Folger Shakespeare Library under the shelfmark V.b.26, is a manuscript of over two hundred pages of rituals and incantations for summoning spirits. This manuscript has been discussed above in the section about the Mercurii, and its quick passage from him to Raphael indicates that it might not have held his interest.[79]

In 1834 George's interest swelled again, as a manuscript from the Cleveland Public Library, *The Key of King Solomon*, indicates. The title page notes that the work was translated from French for George, and gives its date of completion as the 17th of July, 1834. The handwriting and the authorship mark show it to be the work of Frederick Hockley. Indeed, this is the only direct link between the Grahams and Hockley, a famous spiritualist, magician, and collector in his own right. After the book's completion, its ownership remains a mystery. It seems likely to have entered George's library, but even this cannot be certain, given the Grahams' rapidly-changing financial state. It vanishes for a century and a half, until December 20, 1998, when it was purchased from Charles Canfield Brown Books in Jersey City by the Cleveland Public Library. The bookstore went out of business years ago, and any

---

78. Cicero, "Gathering and Bursting of the Astrological Tumour."

79. Harms, Clark, and Peterson, *The Book of Oberon: A Sourcebook of Elizabethan Magic*; Campbell, *A Book of the Offices of Spirits*.

documentation on its previous history has been lost.[80]

What book of magic might have excited George's interest enough for him to commission his own copy? Many modern readers are familiar with the *Key of Solomon* through the edition edited by MacGregor Mathers in 1889, but this title is but one of an immense and variegated collection of works with multiple textual traditions, deriving ultimately from the Greek work entitled *Hygromanteia*. Overall, most editions of the *Key* are dedicated to the summoning of spirits and the creation of talismans, although the exact order and contents of the book can vary considerably. George's version maps most closely to a variant attributed to a pseudonymous copyist, "Armadel," as part of a transmission chain in which the Greek was translated into Latin, then into Italian by Abraham Colorni, and then into French. Other manuscripts in this group include British Library Lansdowne MS. 1202, Houghton Library MS. 554, Wellcome Institute MS. 4660, and Bibilothéque Nationale MS. 2349.[81]

Yet this work has several differences from Lansdowne MS. 1202, to which I have conducted a comparison. Much material has been excised, especially that relating to the preparations—sadly, including the purification for one's dog as a companion—and the spheres of influence of the spirits. The prayers and conjurations have been completely rewritten. The book includes a lengthy list of spirits similar to those in the infamous *Grimorium Verum*; the best parallel to this work, as Joseph Peterson observes, is in a series of Continental manuscripts dating back to the seventeenth century. The talismans are unusual, both visually and in the curious bloods and parchments needed to make them from creatures ranging from tigers to ducks. Finally, we have the insertion of long passages regarding the proper relationship between the master and his companions.[82]

---

80. Cleveland Public Library BF1601 C5313 1834. A facsimile of the work is in the works at Aeon Sophia Press. On Hockley, see Hockley, Hamill, and Gilbert, *The Rosicrucian Seer: Magical Writings of Frederick Hockley*.

81. Marathakis, *The Magical Treatise of Solomon or Hygromanteia Also Called the Apotelesmatike Pragmateia, Epistle to Rehoboam, Solomonike*; Mathiesen, "The Key of Solomon: Toward a Typology of Manuscripts."

82. Peterson, *Secrets of Solomon: A Witch's Handbook from the Trial Records of the Venetian Inquisition*, xix, 54–64.

While working for the bookseller John Denley, Hockley copied several manuscripts of a *Clavis* obtained from the estate of Ebenezer Sibly for clients, but this was a different textual tradition from the one appearing in Graham's manuscript. Based upon the material included in his *Complete Book of Magic Science*, he likely also had access to a manuscript from another Clavis tradition attributed to Pietro Mora. To complete his commission for George, Hockley chose a different manuscript from an entirely different tradition, one which had not been translated into English, as a basis for the work. Further, given that one of his favourite pastimes was copying manuscripts for himself, it is unusual that no copy of this work has turned up in the listings for his library.[83]

What was the motive for this work being copied? One striking aspect of the manuscript are the passages inserted regarding the relationship between a master and his students. This might suggest that George Graham considered starting a magical order of some sort, as Francis Barrett attempted earlier. If so, no further evidence of such an organization has ever been found. At any rate, George had a new project in mind soon after he received the manuscript.

"Aerostation Presents Such Tempting Facilities": The Aeronautical Association

In 1836, George advertised that he, a "veteran," was available "to tavern, inn, and pleasure ground proprieters" for ascents. He also acknowledged that "Mrs. Graham will also ascend, if preferred," and indeed this seems to have been the preference. That year, Margaret made repeated ascents from the Flora Tea Gardens in Bayswater. With her usual reports to the papers, she made a surprising statement: "I have no hesitation in boldly affirming that with a balloon of sufficient dimensions I would willingly pilot 30 persons through the 'realms of air,' and bring them

---

83. Sibley, *Solomon's Clavis, or Key to Unlock the Mysteries of Magic*; Sibley, Hockley, and Peterson, *The Clavis or Key to the Magic of Solomon*; Hockley, *A Complete Book of Magic Science*.

all safely to terra firma."[84]

This was more than idle speculation on her part. The next year saw the debut of a new Aeronautical Association, in which George was to play an important part. Based upon the prospectus, the purpose was "for promoting geographic surveys of some of the remaining undiscovered tracts of the globe, in an aeronautic machine or balloon of larger dimensions than any yet constructed, the first attempt to be directed to the unexplored regions of Africa."[85]

The program described was certainly audacious, especially in the design of the balloon. George's proposal was for two balloons, one nested inside each other, in order to prevent the risk of puncture. He also sought to create a steering apparatus, whereby the aeronauts might ascend and descend, or pass between currents of air, with the utmost ease. If a descent became necessary, multiple grapnels would be on board, so that the loss of one would not cause difficulty. If the balloon did lose its gas, the passengers could easily inflate it with hot air. In the event of a descent into the ocean, such as that the Grahams made near Plymouth, the wicker enclosure that served as the basket could "be easily converted by means of a covering, impervious to water, into a perfect sea-boat." This would also be equipped with a wide variety of devices, including barometers, chronometers, sextants, compasses, telescopes, and a brace of pistols and fowling piece for each participant.[86]

The balloon was to be constructed in London, at a location in which an entrance fee could be charged. A similar fee would be assessed to an audience at each of three inflations at London, as well as a dozen ascents in the countryside of England, Scotland, and Ireland, and possibly a dozen more on the continent. After this, a crew of six, including a draughtsman, a surgeon/naturalist, a geographer, and George himself, would fly low over the unexplored parts of Africa, attempting to gain as much information from their flights as possible.

The company was set up to sell four thousand shares of stock, with the goal of

---

84. "Balloon Ascent—To Tavern, Inn, or Pleasure Ground Proprietors"; "Ascent of Mrs. Graham in a Balloon from Bayswater."

85. Aeronautical Association, "Prospectus of the Aeronautic Association."

86. Aeronautical Association.

taking a minute aerial survey of the interior of Africa,
for which purpose aerostation presents such tempting
facilities, especially with the assistance of the trade winds,
blowing constantly in one direction, so as to render the
steering of the balloon a matter of infinite ease—or rather to do away with the necessity of steering at all![87]

George would have been responsible for piloting the balloon. Given
that he had considerable difficulties with inflating a balloon to ascend
with two people for a few hours, the goal of ascending with dozens for a
long trip across another continent seems foolhardy in the extreme. The
solution was for George to invent further devices to aid in aeronautical
endeavours. One of these would allow a balloon "to go to what point
he pleases, and also to remain up any length of time without loss of
either gas or ballast," while another would allow the balloon to hover
"about 30 feet from the ground" until someone pulled it in, allowing
it to land "50 or 100 times a day without inconvenience to those seated
in the car, or without any loss of gas."[88]

The plan was to be based upon the purchase of 4,000 shares, at a
price of £2 each, made available at the branches of the London and
Westminster Bank. The prospectus promised that only half need be
paid up front, with repayment of the amount to be made within a
year, and additional profits to be paid out therefrom, based upon fees
assessed from audiences and the sales of publications upon the return.
In the end, the plan was doomed to failure; as described in *Chambers's
Journal*, "there were four thousand shares of two pounds each, half to
be paid *on application*, and the public somehow forgot to apply." One of
Margaret's ascents from the Flora Tea Gardens might have encouraged
such forced amnesia.[89]

---

87. "The Air Hath Bubbles."

88. Graham, "Ascent of Mrs. Graham and Captain Currie: Mrs. Graham's Statement";
Graham, "Ascent of Mrs. Graham and Captain Currie of the Guards from Saddler's
Wells Theatre: Mrs. Graham's Statement."

89. Aeronautical Association, "Prospectus of the Aeronautic Association"; "Ballooning,
As It Is Hoped to Be," 643.

Danger, Death, and the Law: The Grahams and Ballooning, 1836–1851

Acceding to the request of William VIII, Duke of Brunswick, Margaret ascended with him from Bayswater on August 22, 1836. Most of the flight went smoothly, but the descent was too rapid. The duke escaped without incident, but the balloon rose again after his departure from the basket, and Margaret fell from almost one hundred feet, leaving her unconscious. She was conducted to a nearby farm in Doddinghurst, with injuries to the head and spine. Although George spoke bravely of her impending recovery, physicians feared she would not survive. Margaret pulled through, although she had bouts of confusion in which she believed she was still in the balloon even a week afterward. Sadly, she had been pregnant at the time, and the child was lost.[90]

We also see the first mention of the Graham children in the coverage of this tragedy. With George at Doddinghurst, the couple's children remained in London, left to their own devices. One burned her face in a candle, and another went to the Middlesex Hospital after a fall. The oldest son, Harvey, about twelve years old, was taken to Doddinghurst to see her in hope that it might help with her recovery.[91]

Margaret's own account of her fall became a topic of some controversy. She maintained that she had fallen from nearly a thousand feet, with her billowing silk pelisse acting as a parachute to slow her descent. This met with some skepticism from some unnamed aeronauts, and the eyewitness accounts indicate instead that she fell no farther than one hundred and fifty feet, turning over in the air repeatedly before impact.[92]

George Graham took over for the rest of the season, making ascents at Hinckley, Manchester, and Bath, although problems with inflating

---

90. A Correspondent, "Aerostation Extraordinary: Ascent of the Duke of Brunswick and Mrs. Graham"; A Correspondent, "Mrs. Graham," August 25, 1836; A Correspondent, "Mrs. Graham," August 26, 1836; "The Late Ascent of the Duke of Brunswick and Mrs. Graham—Melancholy Accident to the Latter, and Loss of the Balloon"; "Mrs. Graham (From the Standard)."

91. A Correspondent, "Mrs. Graham," August 26, 1836.

92. A Correspondent, "Mrs. Graham's Accident"; "Mrs. Graham's Balloon Ascent with the Duke of Brunswick."

the balloon plagued him. His first attempt to ascend at Manchester, on September 12, was unsuccessful. The crowd vented its frustration on the proprietor, manhandling him and tossing his boot about the crowd while local ruffians threatened to blow up the balloon. Fortunately, George was able to stage a successful flight the following day. Leicester was less lucky when he attempted a flight on October 5. According to local accounts, he agreed to satisfy his audience the next day, waited until dark, packed up his balloon, and left town. In the morning, the town rioted, burning down everything in the area and throwing objects at the mayor and constables when they tried to restore order. A man lost his eyesight when rowdies splashed a cask of vitriol in his face during the fracas.[93]

The following year saw both Grahams ready to fly again, with a lighter and involuntary passenger. In their ascent at the Surrey Zoological Gardens on May 15, 1837, they brought along a monkey named Signor Jacopo. Once aloft, the unhappy Signor was ejected while attached to a parachute, luckily making his way to the ground unharmed. The balloon sustained damage that night, but the hard work of thirty seamstresses made it fit to repeat the same trip the next day. Signor Jacopo made a number of other parachute jumps in the following year.[94]

This year also had its minor accidents. George Graham made an ascent at Kingstown, ending in a splashdown in the ocean. Having learned his lesson from Plymouth, he had arranged for a steamship to be on hand to pick him up. He had sent it too far north, however, and it had to chase down the balloon. On one ascent at Lord's Cricket Ground, intended to demonstrate a new parachute design, the balloon with Margaret and her business partner Mr. Adams destroyed a chimney and became entangled in trees. A flight from the Botanical Gardens at Sheffield went off without a hitch, but on its descent, three men attempting to assist were exposed to the gas inside and became unconscious. Finally, while descending near Reigate, Margaret was thrown out of the car and rolled down a steep incline, causing her to

---

93. "Mr. Graham's Balloon Ascent"; "[Untitled]," October 10, 1836.

94. "Surrey Zoological-Gardens"; "Mrs. Graham's Balloon"; "Second Balloon Ascent from Leeds." Another monkey passed away in October, when it fell out of the balloon and no parachute deployed. "Mrs. Graham's Balloon Ascent," October 14, 1837.

be laid up for a while.[95]

By this time, the Grahams had become subjects of parody for more than their astrological interests, as one piece originally published in the *New Monthly Magazine* demonstrates. One Mr. Minnow endeavours to ascend with Mrs. Graham, at which time he will release a novel parachute of his own design. He originally intends to place his infant in the device, but his wife convinces him that a bushel of oysters would be more appropriate. Once a journal editor who accompanies them tosses his latest issue over the side, the balloon bounds into the sky. From their lofty perch, the passengers observe bad sixpences far below. The parachute is dropped without a hitch, but when two gypsies bear back the bushel, only the oyster's shells remain.[96]

All eyes in Britain were on the coronation of Queen Victoria on June 28, 1838. Accepting a government invitation to participate in the festivities, Margaret took her ascent from the Green Park with Captain Currie. The balloon gained considerable height, but the lack of wind made it difficult to find a safe descent. Eventually they floated down near Marylebone Lane, where Currie tossed out a line and called upon the assembled crowd to hold on. They pulled instead, catching the netting on part of a nearby structure's parapet. These pieces struck two people: a servant named Mr. Williams, who was taken to Middlesex Hospital, and John Fley, who was killed by the impact.[97]

Margaret was brought to Mr. Fley's inquest on the 19th of the next month. She gave her account of the accident, on which there was apparently much disagreement. She denied a claim that the balloon had already struck a building before the accident, leaving behind three or four yards of the fabric. Nonetheless, when a witness produced the torn fabric, she had to admit it was hers. She then produced a piece of balloon supposed to belong to one-time ally and current rival balloonist Charles Green. Green, who was present, declared that it was not his, and having been sworn in by acclamation, pronounced the silk of the Graham's balloon too flimsy to be safe. The verdict was an accidental

95. "Mr. Graham—Ascent of the Royal Victoria Balloon at Kingstown"; "[Untitled]," September 13, 1837; "Balloon Ascent at Sheffield"; "Balloons."

96. Browning, "Some Account of the Last Parachute."

97. A Correspondent, "Mrs. Graham's Balloon."

death, with a penalty of twenty shillings assessed against the Grahams.[98]

By this time, the Royal Victoria Balloon was no more. The balloon was in miserable shape, if a writer for *The Satirist* was to be believed; he described it as a "tattered piece of silk" and compared it to an "extensive silk cullender." Margaret had attempted an ascent from the grounds of the Royal Standard tavern, but the wind became so violent that a hole was rent in the side. A constable pulled Margaret out, and her partner John Adams leapt in and commanded the ropes to be released, beginning his first solo flight without preparation. The balloon careened just above buildings, impacting them on occasion, before a garden pole caught it and cut another hole in the side. The balloon crashed into a nearby field, where it was set upon by a mob. Mr. Adams seems to have acquired the luck of the Grahams and was unharmed, but the balloon was a total loss.[99]

The rest of the ballooning season was not much better. Upon the Grahams landing after an ascent at Welchpool, a local labourer cut a wide swath in their balloon with a scythe, thinking the balloonists were within. An attempted inflation at the Birmingham Gas Company went long, enough that the crowd pelted the aeronauts and balloons with brickbats, injuring the craft considerably, before they scattered due to rain.[100]

We hear of no more ascents from the Grahams for a while, but in 1840, they were back in court for a landlord-tenant dispute. John Margetts was believed to own a house at No. 1 Upper John-street, Golden-square, having taken possession of it in 1825. Margetts seems to have been a desultory landlord at best. He utilized a former tenant as the rent-collector for a time, and most of the other tenants left after that individual no longer did so. A renter in the attic refused to acknowledge Margetts as the proper agent, eventually leaving after the landlord removed the windows in his apartment, paid his children to conduct him elsewhere, and had him imprisoned. Margetts' ownership of the property came to be challenged by Jorden Chadwick, who claimed that a forebear, Sir Andrew Chadwick, had once possessed a great deal of

---

98. "Adjourned Inquest on the Body of Mr. John Fley."

99. "[Untitled]," July 22, 1838; "Ascent and Destruction of the Victoria Balloon."

100. "Balloon Ascent at Welchpool"; A Correspondent, "Mrs. Graham's Balloon."

property in the area, including the house in question. Jorden lived next door to George Graham on Poland Street, but George seems to have been unimpressed with him, as later events were to prove.

The situation escalated until an altercation at 1 Upper John-street, in which seventeen men on Margetts' behalf stood against those employed by the Chadwicks. This matter seems to have brought to the Marlborough Street Magistrates Court, at which statements were taken from both George and Margaret Graham.

This was not the end of the matter. At about 12:30 PM on August 5, 1840, George noticed a cart pulled up in front of Mr. Chadwick's, unloading a great quantity of windows, doors, cupboards, shutters, and other items. Why Mr. Graham concluded that these came from the house on Upper-John Street is unknown, but within half an hour he had brought some men to that address, who stood in the door when the cart arrived there for a second trip, along with Chadwick and his son, also named Jorden. The two Jordens could not obtain access to the house, so they drove the cart away.

Margetts brought charges against the father and son, and George Graham was called to testify at the Old Court on August 24, 1840. He attested that he had spoken on several occasions with the elder Chadwick, who admitted on multiple occasions that he had no true legal claim to the property. In fact, Chadwick had gone to the church to have the records altered, so that he could forge a pedigree claiming he was a descendent of Sir Andrew Chadwick, although that worthy in fact belonged to another branch of the family. A solicitor for the defendants insisted that they did have a rentcharge that likely extended to the land in question, and the Chadwicks were released.[101]

Even if this trial was not the impetus, it heralded the Graham's departure from Poland Street. The first census from which we have good information about the Grahams is that of 1841, which finds them at King's Street in St. James Parish. We see a full household for the Grahams, with six children resident at the time, down to Edward at a year and a half. Harvey and Rosa are no longer in residence. Given that fifteen other names occupied the same address, it is likely that this was

---

101. Great Britain Central Criminal Court and Buckler, *Central Criminal Court. Minutes of Evidence*, 12:709–17.

a rooming house of some sort. The family must have lived on Gray's Inn Road for a brief time between Poland Street and this address, as the road is noted as Edward's birthplace in the 1851 and 1861 census.

Little is heard of the Grahams in the press for a decade. It seems that they lived in the parish of Westminster for a brief time, as the 1851 and 1861 censuses indicate this was the birthplace of their youngest daughter, Margaret. They did make their way back to 41 Poland-street, however, for a single notable event in 1845. Workmen had demolished an adjoining house, but no work had been put into shoring up the shared wall. At about half past midnight, the Grahams heard wooden beams crack and the familiar sound of bricks falling off a chimney. The family ran to the street in their night-clothes, only to see a substantial portion of the wall and staircase topple to the ground. All members of the family were fine, but much of their furniture was destroyed.[102]

In 1850, Mrs. Graham was back to ballooning. She brought her children on her flights, ascending with her son Harvey and three of her daughters from the Flora Tea Gardens in Bayswater. These flights were conducted in a new balloon. Mrs. Graham claimed that she and her daughters had sewn it themselves, and it was valued at £300. Sadly, this balloon would soon fall victim to the luck of the Grahams—but not before Margaret's most notable achievement.[103]

On August 7, Margaret became the first British woman to make a solo ascent at night, from the Cremorne Gardens. According to her account, the combination of the cold temperatures and the damp balloon made the ascent quite difficult, with the balloon barely clearing the trees near the field. She had some difficulty with descending, given the lack of anyone nearby to assist with the grapnels. Eventually one caught on a ditch nearby, and Margaret Graham held on for half an hour as the basket turned over repeatedly. The local constabulary finally arrived, but as Mrs. Graham was examining the balloon, someone with a lantern came up behind her, igniting the balloon and burning Margaret's face and clothing. The Grahams made the best of a bad situation by displaying the remains of the balloon at the Gardens and seeking a collection to compensate for their losses. Margaret acquired

---

102. "Fall of a House."

103. "Mrs. Graham's Balloon Ascent," August 4, 1850; "Vauxhall."

a new balloon and was ascending shortly thereafter.[104]

By March of 1851, the Grahams had moved to 15 Liverpool-street, Walworth. The household included George and Margaret, along with their children Alice, Lydia, Frances, Jane, Edward and Margaret, and a 67-year-old widow named Mary Dean. All of them except for Ms. Dean list their profession as "aeronaut." It was this year that they would make their most infamous ascent.

'A Ball of an Inflammatory Nature': The Grahams and the Great Exhibition

In 1851, the Great Exhibition brought millions of visitors, many from other countries, to witness the ingenuity of the British Empire in the fields of technology, science, and art. Despite the structuring of the exhibit space and categories to privilege the British Empire, many Londoners also saw the influx of foreign ideas, technologies, and individuals as disconcerting, or even a potential threat.[105]

Sadly, George became caught up in the speculation, as evidenced by a letter sent to the Home Secretary, Sir George Grey, on April 8 of that year. He relayed reports from his immigrant friends suggesting that a massive uprising was in the works, with the chief instigators being Catholic priests, French dissidents, and hundreds of thousands of Irish. Their technology was impressive:

> I have had described to me some of the missiles intended to be used, one of them termed Crabs, one pound Balls with spikes protruding from them in every direction, in order to lame the horses or men, either police or troops that may be engaged against them; also another is a tube in form of a cane, 3 or 4 are appointed to each leading thoroughfare, the tube contains a ball of an inflammatory nature, this is unobserved blown by the Breath into the various shops towards the evening

---

104. "Mrs. Graham and Her Balloon," August 10, 1850; "Destruction of Mrs. Graham's Balloon"; "Cremorne Gardens"; "Vauxhall."

105. Sperling, "'Wot Is to Be': The Visual Construction of Empire at the Crystal Palace Exhibition, London, 1851."

of the day of the intended outbreak these Balls explode at a given time and the places will be instantly in a Blaze, at the same time the Lodging houses in the various parts of the Metropolis will also be in Flames and any Man, Woman, or Boy remaining out to give notice to the Fire Engines offices, will be instantly stabd, at this time of Consternation and Alarm, it is intended to make an attempt when the Bank of England for which they calculate about 20,000 Men will be sufficient, on Buckingham Palace, and other Buildings...[106]

George had the solution, however:

I have invented what may most properly be termed a Street Shield for Policeman; or any troops in reserve; it is calculated to oppose by means of [it] from 30 to 50 men with the aid of this shield against a force of 50,000 who, with whatever weapons they possess, cannot injure one of the 50 Police coming against them; at the same time those 50 will be enabled to carry destruction before them amongst the multitude, but with the aid of my Eldest Son who is a cabinet maker and for whom I have sent for from Bath from the express purpose (where he is settled with a young Wife and 3 Children) I have completed a perfect model to scale...[107]

Grey passed on the letter to Sir Richard Mayne, and a police officer was sent to speak with Mr. Graham. He found George studying the coloring of stained glass, and he judged him to "be a respectable man, although in needy circumstances." George had left a model of his street shield with the police and wanted to know if it would be adopted. The officer was able to trace his accusations of the vast conspiracy to some of his associates, but he could find no further proof.[108]

The same officer met with the Grahams on May 9 to let them know the decision of the Commissioner—presumably, to decline his offer. He informed the officer that a large quantity of shavings had been placed beneath the Exhibition Hall, to be ignited with matches at an

---

106. Letter from George Graham to Sir George Gray, April 8, 1851, MEPO 2/43.

107. Ibid.

108. Letter to Sir George Gray, April 15, 1851, National Archives at Kew, MEPO 2/43.

appropriate moment, killing many people. This likewise seems to have been void of evidence.[109]

As it turned out, the greatest threats to life and limb at the Exhibition were not foreign agents, but the Grahams themselves. In conjunction with the festivities, they launched a balloon from Batty's Hippodrome, a site for equestrian events on the Kensington road near the Crystal Palace. Shortly after 6 PM that day, with hundreds of observers watching, George and Margaret ascended slowly into the air. No sooner had they done so then the balloon struck the flagstaff on top of the Hippodrome, ripping a hole in the side. The Grahams cast out their grappling hooks, to no avail, and the balloon drifted north. Their oldest son Harvey followed on the ground, but he could only watch helplessly as the balloon floated toward the Crystal Palace.

> ... great fears were entertained for the safety of a portion of that building, and of those who were inside. The police report that at the time alluded to there were between 35,000 and 40,000 persons in the building, some little departure having taken place from the ordinary time of 'clearing out.' In a few minutes the balloon was directly over the transept of the Exhibition. The aeronauts, seeing the fearfully dangerous position in which they were placed, let out the whole of their ballast on to the roof of the palace. At this time the grappling irons were within a few feet of the summit of the transept, and if a hold had been obtained, a vast mass of the building must have been torn away.[110]

Fortunately, only the flagstaffs atop the Crystal Palace suffered damage. Careening across the sky, the balloon drifted to Arlington Street, where the grappling irons caught the parapet of the house of Colonel North, dragging it up until it smashed through the roof. As the flight continued, a number of chimneys on Arlington Street and Park Place also were damaged, as were the roofs on which they fell. Finally, the car became trapped abruptly between two houses, causing the Grahams to be thrown out onto the roof.

Upon arriving, the police took the Grahams to the home of a local

---

109. Letter to Sir George Gray, May 10, 1851, National Archives at Kew, MEPO 2/43.

110. "Disastrous Accident to a Balloon."

physician for treatment. Margaret had received several bruises on the back and a cut on her hand, while George had sustained serious cuts and a broken clavicle and sternum. A police enquiry into the accident revealed that "the silk of which [the balloon] was composed appeared to be in a very decayed state," leading to its breakage when placed under stress. [111]

Margaret realized that this would lead to considerable ill press, so she released the following bold and erroneous statement to the *Times*:

> With regard to any accidents that have occurred to myself in my numerous ascents, I have only confidently to declare that I had met with no more than the most experienced aeronaut of the day—all being liable to mischances, particularly upon such a boisterous day as yesterday. [112]

This did little to avail against the bad press. One story from the time lays out what could have occurred if the Grahams had not been lucky:

> Had the balloon struck the transept of the Exhibition, which it barely cleared, and sent the glass and iron work showering down on the people below, a panic would have seized the visitors, the consequences of which might have been most disastrous. If it had fallen and dragged in Rotten row amongst the horses, many serious accidents would most certainly have occurred; or if it had swept Piccadilly amongst the throng of omnibuses, carriages, and crowds of passengers. And had but a few more cubic feet of gas escaped, one of these events must have been the result. [113]

It ended with a comparison to the damage of a cannon being discharged into homes in London.

The Lady's Magazine was not sympathetic. "[Mrs. Graham] may be the most adroit of aeronauts," it stated, "and her many hair-breadth escapes might favour that belief, could we get rid of the fact that they are *many*." It noted that Margaret had herself admitted that high winds

111. "Disastrous Accident to a Balloon"; "The Late Balloon Accident"; "The Hippodrome Balloon Accident."

112. "The Hippodrome Balloon Accident."

113. "Balls and Balloons."

were blowing at the time, and that she had discouraged a passenger from ascending with her due to the danger. The article ends with a tongue in cheek suggestion that a pension be paid to Margaret to ensure her retirement.[114]

The final irony was George's own display as part of the Great Exhibition in the Crystal Palace: a steering device for balloons of his own invention. This curious contraption had four fan-like wings, comparable to those of a dragonfly, that extended from the sides of the balloon's car. It consisted of four small sails attached to the ring, like the wings of a dragonfly. Ropes led down to the basket which a person in the basket could presumably use to change direction. George claimed that these allowed the balloon "to ascend or descend, move onward or backward, without the loss of either gas or ballast." Given their small size with respect to the balloon itself, which would catch most of any prevailing wind, it is unlikely that any of these would have been successful. We have no indication that this device was ever field tested or added to one of the Grahams' balloons, if only to make the crowd at the Exhibition more at ease.[115]

'A Female Supported a Husband': Margaret Takes Up Alchemy

Far from the scrutiny of the press and public, Margaret had taken up an interest in alchemy. If George's letter to John Denley of 1833 seeking an alchemical partner implies any qualms about bringing Margaret into the study, these had been dismissed by mid-century.

One of the alchemical notebooks at the Wellcome, MS. 2550, bears Margaret's signature at several different points, the first being a receipt dating to September 29, 1851. Further notes in the book appear with dates up to October 5, 1856. Margaret seems to have been interested enough in to compile at least two treatises dealing with the transmutation of metals, the elixir of life, and other alchemical issues. She

---

114. "Mrs. Graham and Her Balloon," August 10, 1850.

115. Great Exhibition, Great Britain., and Commissioners for the Exhibition of 1851, "Official Catalogue of the Great Exhibition of the Works of Industry of All Nations, 1851 ...", "Apparatus for Steering Balloons."

copied wholesale George Ripley's seventeenth-century 'Bosome-Book,' and it is quite likely that the other treatises therein have parallels elsewhere.

As with George, no notebooks indicating experimental activity have been passed on to us, with most of the information being recipes apparently copied from other works. We can only hope that some of these might survive, as Margaret's known manuscripts are in a deplorable state.

The notebook also includes other insights into the Grahams, including their religious beliefs. An article, supposedly copied from the *Bucks Advertiser* in March 1853, relates the predictions of three reverends regarding the end of the world in fourteen years, the status of Napoleon III as one of the heads of the Beast of the Apocalypse, and the possible destruction by Rome through volcanoes due to the sins of the Catholic Church. Also present among these notebooks were an enumeration of the US population broken down by race, as well as slave and free statuses, and a recipe for ginger beer.

Two additions to George Graham's 1824 notebook, breaking up his careful sectioning, may also date to this period of experimentation. One of these describes how one might create glass of different colours, which parallels the statements of the police inspector of George's new interest circa 1850. The other brief treatise is an alchemical operation 'copied from an old manuscript' that describes the creation of silver from base metals. Given its title, 'A Process by which a Female supported a Husband and a worshipful Household in character of a Gentlewoman,' one is tempted to ascribe the addition of this work to Margaret, suggesting she might have taken over the book for herself.

'The Principal Bolts': Last Flights, 1852–53, and the End of the Grahams

The fiasco at the Exhibition was not the end of the Grahams' ascents. On July 7, 1852, Margaret attempted to take two ladies with her in an ascent from the Rosemary Branch Tavern at Hoxton. As they were attaching the car, however, the balloon freed itself and flew up into the air without passengers. It turned over in the air and exploded into small pieces of silk. In 1853, she made a series of ascents in Dublin. In one flight with a Mr. Kennedy, the balloon ascended little, flying through the town and knocking into buildings before becoming attached to one, leaving Mrs. Graham badly bruised.[116]

The Grahams were perhaps the most notable balloonists to experience such troubles, but they were hardly the only ones. The year 1850 alone saw two deaths, one when an airship piloted by Hugh Bell killed a man with its grapnel, and another when George Gale's balloon car overturned in an ascent in Bordeaux. In 1852, one James Goulston, flying as Giuseppe Lunardi, became entangled in the ropes beneath his balloon and was dashed against the wall. None of these made a strong argument for the safety of either passenger or spectator of balloons. Even those who were not injured in person could be injured financially. One W. L. Adams, a farmer near the city, complained as to great damage to his crops as people ran through his fields in pursuit of a balloon. He felt he had little recourse, as "when there's damage, the principal bolts, and the only answer I can get is, 'Oh, it belongs to Mr. Green or Mrs. Graham,' and if I detain it I am liable for damage, for it has to go up to-morrow." Press coverage decreasd, and the wholescale lionization of the aeronauts dropped off. As it turned out, this was the perfect time for the career of the Grahams to end.[117]

On November 3, 1853, the case "Graham v Beardsley" was brought to the Manchester County Court. A Mr. Beardsley had engaged Margaret to lift off from Pomona Gardens, but after repeated attempts, she was not able to do so. After one unsuccessful inflation, the balloon would not deflate, and Mrs. Graham took the unusual step of having cuts made in the side of the balloon to do so. Beardsley offered her only partial payment, and Margaret brought him to court for the rest. The prosecution's argument was that the coal gas provided was of insufficient buoyancy, but the opinion of the defence—and the reporter covering the trial—was that the balloon was "certainly in a sadly patched and dilapidated state." Nonetheless, after Judge Brandt examined the bal-

116. "Loss of Mrs. Graham's Balloon"; "Dangerous Balloon Ascent"; Saunders, "The Balloon—Dangerous Descent"; "Rotundo Gardens—Mrs. Graham's Ascent."

117. Adams, "A Claimant for Protection."

loon, he ruled in Margaret's favour, granting her £30.[118]

The final reference to the Graham's long careers as balloonist was, appropriately enough, a lawsuit in 1854. Margaret had been offered 30£ to ascend from Cheveley Park at Newmarket, but she had not made the ascent due to lack of sufficient gas. Instead, she had demonstrated George's "steering apparatus" and made a speech apologizing to the crowd. The proprietor of the festival to which it was attached, Mr. Fairlie, refused to pay her the amount. The court found that he had not fulfilled the terms of the contract and instructed him to pay Mrs. Graham the amount in full.[119]

The last glimpse we have of the Graham household comes through the census of 1861, which finds George and Margaret living at 38 East-cheap with Jane, Edward W., and the younger Margaret. The two elder Grahams both continue to list themselves as "aeronauts," but Jane and young Margaret are cap milliners, perhaps an outgrowth of the Graham's clothing / astrology shop, and Edward is enrolled in the Surrey Volunteers. The 7th Surrey Volunteers were a new unit at the time, having paraded for the first time only the previous year.

By the time of their deaths, George and Margaret had sunk back into obscurity. George died on April 24, 1867, and he was buried six days later in the Abney Park Cemetery. The death record indicates he was living in Hackney at the time. A Margaret Graham lies hundreds of feet away in the same cemetery; she was born in 1796, which runs against our census data, and was buried on July 2, 1864. Her precise date of death is unknown.

Reflections

Even after years of research, I do not feel that I know them. So much of them is filtered through the writing of their contemporaries, in most cases appealing to a mass audience. Even when the Grahams' voices are heard, they are consciously creating themselves, their objectives, and their characters for that same audience. Even their private books,

---

118. "Mrs. Graham and Her Balloon," November 7, 1853; "The Balloon Failure at the Pomona Gardens."

119. "Court of Exchequer, Friday, Dec. 22."

in the fields of ritual magic and alchemy, fail to reveal their voices. No diary or correspondence that might illuminate their private thoughts has been found, and this element is sorely missed.

What can be said about the Grahams? George and Margaret were certainly brave, as any aeronauts of the time would have to be. They were resilient and resourceful, able to rebound after countless injuries, bankruptcies, and destructions of property. They were able to play to their audience's love of drama, but their occult works display a quiet sincerity, even in the pursuit of topics that are still not considered respectable.

At the same time, however, their conduct while ballooning is seriously troubling. We can easily set aside their pretensions as to scientific attainment, which yielded nothing of worth; certainly, many performers have done so now and at other times. We can also hold them largely harmless for the weather, the unruly crowds, and the inherent danger of their endeavours. If a theme emerges regarding the misfortunes of the Grahams, however, it was the wilful and oft-repeated decision, throughout their careers, to ascend in balloons without enough buoyancy to keep their passengers safely aloft, causing great danger to both themselves and others, leading to injury and death.

Although past authors on ballooning have treated the Grahams, none has dealt with them at any length. This is understandable; we want to see balloonists as a part of the grand innovations that brought about human flight, tying together the Montgolfiers, the Wright Brothers, and Neil Armstrong. The Grahams, with their commercial instincts and their rampant endangerment of others, do not fit this narrative. Nonetheless, they can tell us much about popular attitudes and practices for both ballooning and esotericism—and if further documentation can be recovered, we might learn much more about these fascinating individuals.

Bibliography

A Correspondent. "Aerostation Extraordinary: Ascent of the Duke of Brunswick and Mrs. Graham." *The Times of London.* August 23, 1836.

———. "Mrs. Graham." *The Times of London.* August 25, 1836.

———. "Mrs. Graham." *The Times of London.* August 26, 1836.

———. "Mrs. Graham's Accident." *The Times of London.* August 27, 1836.

———. "Mrs. Graham's Balloon." *The Times of London.* June 30, 1838.

"A Ghost." *The Times of London*, September 23, 1828.

"Account of Mr. Harris' Ascent, and the Fatal Result That Attended It." *Bell's Weekly Messenger*, May 30, 1824.

Adams, W. L. "A Claimant for Protection." *The Times of London.* August 26, 1852.

"Adjourned Inquest on the Body of Mr. John Fley." *The Times of London.* July 20, 1838.

"[Advertisement for the Astronomical and Astrological Society of Great Britain]." *The Athenaeum: Journal of English and Foreign Literature, Science, and the Fine Arts*, May 10, 1834.

Aeronautical Association. "Prospectus of the Aeronautic Association," 1837. National Aerospace Library.

Agrippa von Nettesheim, Heinrich Cornelius, and Petrus. *Henry Cornelius Agrippa, His Fourth Book of Occult Philosophy; Of Geomancie; Magical Elements of Peter de Aban; Astronomical Geomancie; The Nature of Spirits; Arbatel of Magick; The Species or Several Kindes of Magick.* Edited by Robert Turner. London: J. C., 1665.

"Apparatus for Steering Balloons." *Illustrated London News*, June 12, 1852. National Aerospace Library.

"Ascent and Descent of Mr. Graham's Balloon." *The Times of London*, September 13, 1823.

"Ascent and Destruction of the Victoria Balloon." *The Times of London.* July 18, 1838.

"Ascent of Mrs. Graham in a Balloon from Bayswater." *The Times of London.* June 29, 1836.

"Balloon and Parachute." *Essex Herald*, May 19, 1829.

"Balloon Ascent." *Morning Advertiser*, May 12, 1825.

"Balloon Ascent." *The Times of London.* November 18, 1825.

"Balloon Ascent." *The Times of London.* October 22, 1828.

"Balloon Ascent." *Cambridge Chronicle and Journal*, November 7, 1828.

"Balloon Ascent." *The Times of London.* September 5, 1831.

"Balloon Ascent—To Tavern, Inn, or Pleasure Ground Proprietors." *Bell's Life in London and Sporting Chronicle*, April 17, 1836.

"Balloon Ascent at Sheffield." *The Times of London*. October 6, 1837.

"Balloon Ascent at Welchpool." *The Times of London*. October 10, 1838.

"Balloon Ascent from Glocester." *The Times of London*. November 17, 1836.

"Balloon Hoax and Riot." *The Times of London*. August 19, 1823.

"Ballooning, As It Is Hoped to Be." *Chambers's Journal of Popular Literature, Science, and Art*, October 13, 1866.

"Balloons." *The Idler and Breakfast Table Companion*, May 27, 1837.

"Balls and Balloons." *The Cornish Telegraph*, June 27, 1851.

Barrett, Francis. *The Magus, or, Celestial Intelligencer: Being a Complete System of Occult Philosophy in Three Books Containing the Antient and Modern Practice of the Cabalistic Art, Natural and Celestial Magic, &c. ... Exhibiting the Sciences of Natural Magic, Alchymy, or Hermetic Philosophy; Also the Nature, Creation, and Fall of Man...; To Which Is Added Biographia Antiqua, or the Lives of the Most Eminent Philosophers, Magi, &c.: The Whole Illustrated with a Great Variety of Curious Engravings, Magical, and Cabalistical Figures, & C.* London: Printed for Lackington, Allen, and Co ..., 1801.

Brailsford, Dennis. *Bareknuckles: A Social History of Prize-Fighting*. Cambridge [England]: Lutterworth, 1988.

Browning, Henry. "Some Account of the Last Parachute." *The Times of London*. October 7, 1837.

Campbell, Colin. *A Book of the Offices of Spirits, the Occult Virtue of Plants, & Some Rare Magical Charms & Spells, Transcribed by Frederick Hockley from a Sixteenth Century Manuscript on Magic & Necromancy by John Porter*. York Beach, ME: The Teitan Press, 2011.

Cicero. "Gathering and Bursting of the Astrological Tumour," n.d.

Cole, Douglas. "Sigismund Bacstrom's Northwest Coast Drawings and an Account of His Curious Career." *BC Studies*, no. 46 (1980): 61–86.

"Colonel Harvey's Flight." *Bury and Norwich Post, or, Suffolk and Norfolk Telegraph, Essex, Cambridge and Ely Intelligencer*. September 14, 1825.

"Cosmorama to Be Sold." *Morning Advertiser*, February 8, 1831.

"Court of Common Pleas, February 6." *The Times of London*. February 7, 1833.

"Court of Common Pleas, Friday, May 27: Smithies v. England." *Morning Chronicle*, May 28, 1825.

"Court of Exchequer, Friday, Dec. 22." *The Times of London*. December 23, 1854.

"Court of King's-Bench, Westminster, June 8, Maton and Grafton v. Brooker." *The Times of London*, June 9, 1827.

"Cremorne Gardens." *Bell's Life in London and Sporting Chronicle*, August 11, 1850.

"Dangerous Balloon Ascent." *The Times of London*. August 22, 1853.

Davies, Mark. *King of All Balloons: The Adventurous Life of James Sadler, the First English Aeronaut*, 2015.

Davies, Owen. *Witchcraft, Magic and Culture, 1736–1951*. Manchester: Manchester University Press, 1999.

"Death of Mr. Harris the Aeronaut." *The Times of London*. May 27, 1824.

Decker, Ronald, Thierry Depaulis, and Michael Dummett. *A Wicked Pack of Cards: The Origins of the Occult Tarot*. New York: St. Martin's Press, 1996.

Demarest, Marc. "A Trout in the Milk: Plotting the Orphic Circle." Chasing Down Emma: Resolving the Contradictions of, and Filling In the Gaps in, the Life, Work, and World of Emma Hardinge Britten, March 25, 2009. http://ehbritten.blogspot.com/2009/03/trout-in-milk-plotting-orphic-circle.html.

———. "Hypotheses on the Orphic Circle." Chasing Down Emma: Resolving the Contradictions of, and Filling In the Gaps in, the Life, Work, and World of Emma Hardinge Britten, June 2011. www.ehbritten.org/papers/hypotheses_on_the_orphic_circle.pdf.

"Destruction of Mrs. Graham's Balloon." *Illustrated London News*, August 17, 1850.

"Diary of Public Sales, Feb. 27." *Morning Advertiser*, February 28, 1835.

Dimitriadis, Dicta. *Mademoiselle Lenormand: La Reine de la Voyance*. Paris: Perrin, 1990.

"Disastrous Accident to a Balloon." *The Times of London*. June 17, 1851.

Dixon. *The True Prophetic Messenger for 1833: Containing Remarkable Events, Predictions, the Weather, &c. ...* [London]: J. Denley, 1832.

"Fall of a House." *Morning Chronicle*, September 25, 1845.

"Fatal Aeronautic Excursion." *The Observer*, May 31, 1824.

G., G. W., and R. C. S. *The Philosophical Merlin, Being the Translation of a Valuable Manuscript, Formerly in the Possession of Napoleon Bonaparte ... Enabling the Reader to Cast the Nativity of Himself ... Without the Aid of Tables ... or ... Calculations*. London, 1822.

Gansten, Martin. "Placidean Teachings in Early Nineteenth-Century Britain: John Worsdale and Thomas Oxley." In *Astrologies: Plurality and Diversity: The Proceedings of the Eighth Annual Conference of the Sophia Centre for the Study of Cosmology in Culture, University of Wales, Trinity Saint David, 24–25 July 2010.*, edited by Nicholas Campion and Sophia Centre for the Study of Cosmology in Culture. Sophia Centre Press, 2011.

Glatstein, Jeremy. "Sigismund Bacstrom's Alchemical Manuscripts." *Getty Research Journal* 2 (2010): 161–68.

Godwin, Joscelyn. *The Theosophical Enlightenment.* Albany: State University of New York Press, 1994.

Graham, George. "Mr. Graham's Account of His Ascent." *The Times of London.* September 8, 1823.

———. "To the Public." W. Tickner, March 29, 1824. Cuthbert Aeronautical Collections.

Graham, Margaret. "Ascent of Mrs. Graham and Captain Currie: Mrs. Graham's Statement." *The Times of London.* July 23, 1836.

———. "Ascent of Mrs. Graham and Captain Currie of the Guards from Saddler's Wells Theatre: Mrs. Graham's Statement." *The Times of London.* August 5, 1836.

———. *Statement of the Particulars Relative to Mr. Graham's Balloon, Explaining the Cause of Disappointment to the Public, the Heavy Loss He Sustained from It, and an Account of the Receipts and Expenditure.* Norwich: W. Booth, 1825.

"Graham's Balloon." *Westmoreland Gazette,* June 12, 1824.

Great Britain Central Criminal Court, and H. Buckler. *Central Criminal Court. Minutes of Evidence.* Vol. 12. George Herbert, 1840. https://books.google.com/books?id=3a4DAAAAQAAJ.

Great Exhibition, Great Britain., and Commissioners for the Exhibition of 1851. "Official Catalogue of the Great Exhibition of the Works of Industry of All Nations, 1851 ..." London: Spicer Bros., 1851.

Harms, Daniel, James R. Clark, and Joseph H. Peterson. *The Book of Oberon: A Sourcebook of Elizabethan Magic.* Woodbury, MN: Llewellyn, 2015.

Hills, Gordon P. G. "Notes on General Charles Rainsford (1728–1809) and His Rosicrucian Studies as Illustrated by the Rainsford Papers, Add. MSS. Nos. 23,644-23,680 in the British Museum Library." *Transactions of the Metropolitan College, SRIA,* 1922, 7–29.

Hills, Gordon Pettigrew Graham. "Notes on Some Masonic Personalities at the End of the Eighteenth Century." *Ars Quatuor Coronatorum: Transactions of the Quatuor Coronati Lodge No. 2076 London* 25 (1912): 141–64.

Hockley, Frederick. *A Complete Book of Magic Science, Containing the Method of Constraining and Exorcising Spirits to Appearance, the Consecration of Magic Circles, and the Form of a Bond of Spirits, Transcribed from an Ancient Manuscript Grimoire.* York Beach, ME: The Teitan Press, 2008.

Hockley, Frederick, John Hamill, and R. A. Gilbert. *The Rosicrucian Seer: Magical Writings of Frederick Hockley.* York Beach, ME: The Teitan Press, 2009.

Howe, Ellic. *Raphael; or, The Royal Merlin.* London: Arborfield, 1964.

Jackson, Alexander Cosby Fishburn. *Rose Croix: The History of the Ancient and Accepted Rite for England and Wales.* [Hinckley]: Lewis Masonic, 2005.

King, Francis. *The Flying Sorcerer: Being the Magical and Aeronautical Adventures of Francis Barrett, Author of The Magus.* Oxford: Mandrake, 1992.

Knight, Stephen. *Merlin: Knowledge and Power through the Ages.* Ithaca: Cornell University Press, 2009.

"Law Intelligence, Court of King's Bench—Monday, Jan. 9: Air Balloons—Heely v. Graham and Webb." *Morning Advertiser*, January 10, 1826.

Lévi, Éliphas, and Arthur Edward Waite. *The Mysteries of Magic: A Digest of the Writings by Éliphas Lévi.* 2nd ed. London: K. Paul, Trench, Trübner, 1897.

London City Council. *Survey of London.* Vol. 31–32. London: F H W Sheppard, 1963. http://www.british-history.ac.uk/survey-london/vols31-2/pt2/pp268-283.

"Loss of Mrs. Graham's Balloon." *The Times of London.* July 12, 1852.

Marathakis, Ioannis. *The Magical Treatise of Solomon or Hygromanteia Also Called the Apotelesmatike Pragmateia, Epistle to Rehoboam, Solomonike.* Singapore: Golden Hoard Press, 2011.

Mathiesen, Robert. "The Key of Solomon: Toward a Typology of Manuscripts." *Societas Magica Newsletter*, no. 17 (2007): 1, 3–9.

McIntosh, Christopher. *The Rosicrucians: The History, Mythology, and Rituals of an Esoteric Order.* York Beach, Me.: S. Weiser, 1997.

McLean, Adam. "Bacstrom's Rosicrucian Society." *Hermetic Journal*, no. 6 (1979): 25–29.

———. "General Rainsford: An Alchemical and Rosicrucian Enthusiast." *Hermetic Journal*, 1990, 129–34.

"Middlesex Sessions, Thursday, Oct. 28." *The Times of London*, October 29, 1824.

"Most Perilous Balloon Ascent." *Northampton Mercury*, October 25, 1828.

"Mr. Graham—Ascent of the Royal Victoria Balloon at Kingstown." *Freemans Journal*, July 17, 1837.

"Mr. Graham and His Balloon." *The Times of London.* September 6, 1823.

"Mr. Graham and His Balloon." *Birmingham Journal*, October 21, 1826.

"Mr. Graham's Balloon." *Morning Chronicle*, August 19, 1823.

"Mr. Graham's Balloon." *Morning Post*, August 20, 1823.

"Mr. Graham's Balloon." *Morning Chronicle*, September 13, 1823.

"Mr. Graham's Balloon Ascent." *Manchester Guardian*, September 15, 1836.

"Mr. Harris and Miss Stokes." *The Times of London.* May 28, 1824.

"Mrs. Graham and Her Balloon." *The Times of London.* August 10, 1850.

"Mrs. Graham and Her Balloon." *Daily News*, November 7, 1853.

"Mrs. Graham (From the Standard)." *The Times of London.* August 31, 1836.

"Mrs. Graham, the Aeronaut." *The Times of London.* June 30, 1826.

"Mrs. Graham's Balloon." *The Times of London.* May 17, 1837.

"Mrs. Graham's Balloon Ascent." *Gloucestershire Chronicle*, October 14, 1837.

"Mrs. Graham's Balloon Ascent." *Era*, August 4, 1850.

"Mrs. Graham's Balloon Ascent with the Duke of Brunswick." *The Champion*, October 10, 1836.

"Norwich, September 7." *Bury and Norwich Post, or, Suffolk and Norfolk Telegraph, Essex, Cambridge and Ely Intelligencer*. September 7, 1825.

Penny, John. *Up, Up and Away!: An Account of Ballooning in and around Bristol and Bath 1784–1999*. Local History Pamphlets 97. Bristol: Bristol Branch of the Historical Association, 1999.

Peterson, Joseph H., ed. *Secrets of Solomon: A Witch's Handbook from the Trial Records of the Venetian Inquisition*. Kasson, MN: Twilit Grotto, 2018.

"Police: Marlborough-Street." *The Times of London*, October 15, 1824.

"Police: Worship Street." *The Times of London*. July 4, 1826.

Priddle, Robert A. "More Cunning than Folk: An Analysis of Francis Barrett's *The Magus* as Indicative of a Transitional Period of English Magic." Thesis, The University of Ottaowa, 2012.

"Questions Continued: X." *Birmingham Chronicle*, January 25, 1827.

Raphael. *Raphael's Sanctuary of the Astral Art: Or, Elysium of Astrology: Being a Book for the Boudoir, Drawing-Room Table, and Evening Parties, Containing a Complete Geomantic Cabinet Illustrated with Emblematical Pictures of the Twelve Celestial Houses; Also, Spirits of the Earth, Air, Fire, and Water, & c. & c. & C*. London: W.C. Wright, 1834.

———. *The Familiar Astrologer: An Easy Guide to Fate, Destiny, & Foreknowledge, As Well As to the Secret and Wonderful Properties of Nature*. London: Printed for John Bennett, 1832.

Raphael, and Merlinus Anglicus. *The Astrologer of the Nineteenth Century*. London: Knight & Lacey, 1825.

"Rotundo Gardens—Mrs. Graham's Ascent." *Freemans Journal*, August 11, 1853.

"Sale by Auction." *Morning Advertiser*, June 9, 1831.

"Sales by Auction." *Morning Advertiser*, May 13, 1831.

"Sales by Auction." *Morning Advertiser*, May 16, 1831.

"Sales by Auction." *Morning Advertiser*, May 27, 1831.

"Sales by Auction." *Morning Advertiser*, June 16, 1831.

Saunders. "The Balloon—Dangerous Descent." *Cork Constitution*, August 23, 1853.

Schuchard, Marsha K. M. *Freemasonry, Secret Societies, and the Continuity of the Occult Traditions in English Literature*. Ann Arbor: University of Michigan, 1975.

"Second Balloon Ascent from Leeds." *The Leeds Mercury*, October 28, 1837.

Sibley, Ebenezer. *Solomon's Clavis, or Key to Unlock the Mysteries of Magic*. Burbage: Society for Esoteric Endeavour, 2008.

Sibley, Ebenezer, Frederick Hockley, and Joseph H. Peterson. *The Clavis or Key to the Magic of Solomon.* Lake Worth, FL; Newburyport, MA: Ibis Press; Distributed by Red Wheel/Weiser, 2009.

Skinner, Stephen. *Geomancy in Theory and Practice: The Most Complete History of Western Divinatory Geomancy in English.* Singapore: Golden Hoard Press, 2011.

Smith, Digby George. *1813, Leipzig: Napoleon and the Battle of the Nations.* London; Mechanicsburg, PA: Greenhill Books; Stackpole Books, 2001.

Sommers, Susan Mitchell. *The Siblys of London: A Family on the Esoteric Fringes of Georgian England.* New York: Oxford University Press, 2018.

Sperling, Joy. "'Wot Is To Be': The Visual Construction of Empire at the Crystal Palace Exhibition, London, 1851." In *Fear, Loathing, and Victorian Xenophobia,* edited by Marlene Tromp, Maria K. Bachman, and Heidi Kaufman, 181–207. Columbus: Ohio State University Press, 2013.

Stark, W. *A Letter to John Harvey Containing an Examination of a Pamphlet Entitled, "Mrs. Graham's Statement of Facts" Relative to the Balloon.* Norwich: Burks and Kinnebrook, 1825.

Summers, Judith. *Soho: A History of London's Most Colourful Neighbourhood.* London: Bloomsbury, 1989.

Summers, Montague. *Witchcraft and Black Magic.* New York: Causeway Books, 1974.

"Surrey Zoological-Gardens." *The Times of London.* May 16, 1837.

"The Air Hath Bubbles." *Mechanics' Magazine, Museum, Register, Journal, and Gazette,* February 18, 1837.

"The Balloon." *Norwich Mercury,* September 10, 1825.

"The Balloon Failure at the Pomona Gardens." *Manchester Guardian,* November 12, 1853.

The Chemical Society of London. "Proceedings of the Meeting of the Chemical Society." *The Journal of the Chemical Society of London* 17 (1864): 433–53.

"The End of the World." *Northampton Mercury,* November 1, 1828.

"The Grecian Balloon." *The Times of London.* September 7, 1824.

"The Hippodrome Balloon Accident." *The Times of London.* June 19, 1851.

"The Inquest." *The Times of London.* May 27, 1824.

"The Late Ascent of the Duke of Brunswick and Mrs. Graham—Melancholy Accident to the Latter, and Loss of the Balloon." *The Times of London.* August 24, 1836.

"The Late Balloon Accident." *The Times of London.* June 18, 1851.

"The Late Balloon Ascent." *The Times of London.* November 23, 1825.

"The Members of the Mercurii." *The Western Times,* August 19, 1851.

"[Untitled]." *Bell's Life in London and Sporting Chronicle,* August 21, 1825.

66

"[Untitled]." *The Times of London*. June 29, 1826.

"[Untitled]." *Staffordshire Advertiser*, August 1, 1829.

"[Untitled]." *The Times of London*. October 10, 1836.

"[Untitled]." *The Times of London*. September 13, 1837.

"[Untitled]." *The Satirist or the Censor of the TImes*, July 22, 1838.

"Vauxhall." *Bell's Life in London and Sporting Chronicle*, August 25, 1850.

Waite, Arthur Edward. *The Brotherhood of the Rosy Cross, Being Records of the House of the Holy Spirit in Its Inward and Outward History*. New Hyde Park, N.Y.: University Books, 1961.

Wilson, H. C. Bruce. "The Origin of Our Rosicrucian Society." In *The Origins of the Rosicrucian Society in England*, edited by Darcy. Kuntz, 8–22. Austin (Tex.): Golden Dawn Research Trust, 2009.

Wood, Stephen. *Those Terrible Grey Horses: An Illustrated History of the Royal Scots Dragoon Guards*. Oxford: Osprey Pub Co, 2015.

Index

A

Abney Park Cemetery, 56
Adams, John, 22, 28, 44, 46
Adams, W. L., 55
Aeronautical Association, 40–42
Ager, 30
Agrippa, Heinrich Cornelius, 6
Alchemy, 3, 7–11, 32–33, 35–37, 52–53, 57
Algar, John, 31
Ancient Order of Druids, 2
Armadel, 39
Astrologer of the Nineteenth Century (Smith), 11, 23, 37
Astrology, 1, 3, 6, 11–13, 16, 22, 32–38, 45
Astronomical and Astrological Society of Great Britain, 33–35

B

Bacstrom, Sigismund, 8–11, 35
Bailey, Charles, 29
Barrett, Francis, 10–11, 14, 38, 40
Bath, 21, 43, 50
Bath Gas Light & Coke Company, 21
Battle of Leipsic, 5
Batty's Hippodrome, 51–52
Bayswater, 40–41, 43, 48
Beardsley, 55–56
Bell, Hugh, 55
Bell, Shakespear, 24–26
Bennington, 31–32
Birmingham, 28–29, 46
Birmingham Gas Company, 46
Blake, William, 2
Booth, W., 25
"Bosome-Book" (Ripley), 54
Botanical Gardens, Sheffield, 44
Boyes, Henry Cowell, 9, 35
Bramble, 26

Victoria I, Queen, 45

W

W., H., 13–14
Waite, Arthur Edward, 2–3, 8
Walcot St. Swithin, 16
Webb / Webbe, 28
Welchpool, 46
Wellcome Institute, 35, 39, 53
White, Mr., 25
White-Conduit House, Pentonville, 17–18, 28
William Henry White, William Henry, 8
Williams, George, 16
Williams, Hannah, 16
Williams, Margaret, *see* Graham, Margaret
Williams, Mr., 45
William VIII, Duke of Brunswick, 37, 43
Windsor, 31
Winston, James, 2
Wonderful Magical Scrapbook, 33–37
Woulfe, Peter, 9–10

X

Y

Z

"Zadkiel". *See* Morrison, Richard James

Typesetting and cover design by Casey Hickey
Composed in Adobe Garamond Pro by Robert Slimbach

Designer's Note
The task of typesetting historical content often leads to adventures in the history of typography and graphic design. Such an effort will typically illuminate a typeface that corresponds with the time and place of the subject matter.

In this case, several obstacles lead to an anachronistic choice of Adobe Garamond Pro—a typeface based on the work of two sixteenth-century French designers: Claude Garamond and Robert Granjon[1].

England was home to John Baskerville, an eighteenth-century businessman and print innovator whose work influenced many of the prominent nineteenth-century type designers in Italy and France. His deviations from Renaissance letterform conventions—as well as jealousy from professional peers—made him less popular in England.[2] Despite hostile critics, it seems reasonable to expect that another English designer would have picked up where Baskerville left off. Though the influence of his work is apparent, evidence of an English successor to Baskerville is not present in comprehensive sources including *History of Graphic Design* by Philip Meggs and *A Short History of the Printed Word* by Warren Chappell.

Unfortunately, the readily available Baskerville font lacked certain OpenType features that would make it a good choice for body text. After a few other more thematically appropriate typefaces were also found to be lacking features, Adobe Garamond Pro was chosen as a full-featured typeface that both the author and designer could accept, despite the thematic mismatch.

1. *Wikipedia, The Free Encyclopedia*, s.v. "Adobe Garamond," (accessed February 16, 2019), https://en.wikipedia.org/wiki/Garamond#Adobe_Garamond

2. Philip Meggs, *A History of Graphic Design*, 3rd ed. New York: John Wiley & Sons, 1998. 112–115